# Springer Proceedings in Mathematics & Statistics

Volume 67

For further volumes:
http://www.springer.com/series/10533

**Springer Proceedings in Mathematics & Statistics**

This book series features volumes composed of selected contributions from workshops and conferences in all areas of current research in mathematics and statistics, including OR and optimization. In addition to an overall evaluation of the interest, scientific quality, and timeliness of each proposal at the hands of the publisher, individual contributions are all refereed to the high quality standards of leading journals in the field. Thus, this series provides the research community with well-edited, authoritative reports on developments in the most exciting areas of mathematical and statistical research today.

Marcello Delitala · Giulia Ajmone Marsan
Editors

# Managing Complexity, Reducing Perplexity

Modeling Biological Systems

1st Kepler Prize Workshop of the European Academy of Sciences (EURASC), Heidelberg, May 2011

*Editors*
Marcello Delitala
Department of Mathematical Sciences
Politecnico di Torino
Torino
Italy

Giulia Ajmone Marsan
OECD
Paris
France

ISSN 2194-1009
ISBN 978-3-319-03758-5
DOI 10.1007/978-3-319-03759-2
Springer Cham Heidelberg New York Dordrecht London

ISSN 2194-1017 (electronic)
ISBN 978-3-319-03759-2 (eBook)

Library of Congress Control Number: 2014940319

© Springer International Publishing Switzerland 2014
This work is subject to copyright. All rights are reserved by the Publisher, whether the whole or part of the material is concerned, specifically the rights of translation, reprinting, reuse of illustrations, recitation, broadcasting, reproduction on microfilms or in any other physical way, and transmission or information storage and retrieval, electronic adaptation, computer software, or by similar or dissimilar methodology now known or hereafter developed. Exempted from this legal reservation are brief excerpts in connection with reviews or scholarly analysis or material supplied specifically for the purpose of being entered and executed on a computer system, for exclusive use by the purchaser of the work. Duplication of this publication or parts thereof is permitted only under the provisions of the Copyright Law of the Publisher's location, in its current version, and permission for use must always be obtained from Springer. Permissions for use may be obtained through RightsLink at the Copyright Clearance Center. Violations are liable to prosecution under the respective Copyright Law. The use of general descriptive names, registered names, trademarks, service marks, etc. in this publication does not imply, even in the absence of a specific statement, that such names are exempt from the relevant protective laws and regulations and therefore free for general use.
While the advice and information in this book are believed to be true and accurate at the date of publication, neither the authors nor the editors nor the publisher can accept any legal responsibility for any errors or omissions that may be made. The publisher makes no warranty, express or implied, with respect to the material contained herein.

Printed on acid-free paper

Springer is part of Springer Science+Business Media (www.springer.com)

# Foreword

This volume presents contributions of the workshop designed and organized by Giulia Ajmone Marsan (France), Marcello Delitala (Italy), Andrea Picco (Germany), the winners of the Kepler Prize for European Young Scientists (KEYS), established by the European Academy of Sciences (EURASC) (http://www.eurasc.org) in 2010/11.

The first edition of the prize was dedicated to the general topic *Mathematical Modelling and Simulations in Life Sciences* and the winning project had the title *Complex Living Systems: Managing complexity, reducing perplexity*. The general topic was chosen since there is an urgent demand for experts in this field as well as in basic research as in applications; projects require interdisciplinary teams including experts from biology, medicine, biophysics and biochemistry, mathematics, and computational sciences.

The Kepler Prize has the goal of steering the cooperation of highly talented young scientists in Europe interested in research crossing the borders of disciplines and states, thus participating in the building of a European Research Area. The award is granted to an international team of young scientists, selected in an international competition for a workshop covering multidisciplinary topics, planned and organized by the team. It consists of financial and organizational support to run the workshop. Further, the proponents of the winning team are invited to become Kepler fellows of EURASC for 3 years, and are expected to take part in scientific activities of EURASC. The selection criteria include excellence of participants; excellence of the proposal with respect to cross-disciplinarity, knowledge transfer and dialogue with society; geographical heterogeneity. Establishing the Kepler Prize EURASC is in line with the Commission of the European Union, stating

- *Young researchers trained in Europe should be confident that their qualifications will be rewarding for their careers.*
- *European doctoral programmes and further training should meet stringent quality standards, fulfil the needs of both academia and business, and be recognised across Europe.*
- *Researchers at all levels should be trained in cross-disciplinary work and S&T administration, including knowledge transfer and dialogue with society.*

(GREEN PAPER, The European Research Area: New Perspectives. SEC(2007) 412, Brussels, 4.4.2007).

It was a proper choice to select Johannes Kepler as a leading figure for a prize to excellent young scientists who look not only beyond their own discipline, but who also transgress it. Kepler was an outstanding scholar integrating, in particular, natural philosophy, mathematics, astronomy, astrology, optics, and theology. His scientific achievements revolutionized the view of the world. On the occasion of the three hundredth anniversary of Kepler's death, Albert Einstein wrote in the Frankfurter Zeitung (November 9, 1930)*: *It seems that the human mind has first to construct forms independently, before we can find them in things. Kepler's marvelous achievement is a particularly fine example of the truth that knowledge cannot spring from experience alone, but only from the comparison of the inventions of the intellect with observed fact.*

Johannes Kepler is most famous for discovering the laws for the motion of the planets, *discovering a way of bringing order into this chaos* (Einstein, in *). Kepler himself stated in Astronomia Nova (1609): *The chief aim of all investigations of the external world should be to discover the rational order and harmony which has been imposed on it by God and which He revealed to us in the language of mathematics.*

Understanding the dynamics and structures of processes in life sciences is posing challenges, where quite similar a way has to be found to resolve the complexity of data and observations. The prize winning workshop was going by a very demanding title. Indeed, mastering complexity is a crucial task in most human activities, in particular in science. Living systems involve a higher degree of complexity than purely physical-chemical systems. Necessary for reduction of complexity is a well-defined problem definition, and the formulation of a well-defined question to be answered. As Kepler stated, mathematics plays an important role, not just in formulating *rational order and harmony,* but also in making processes in nature computable and predictable. Exploring the universe using advanced instruments and information technology, including mathematical modeling and simulation, leads to answers to crucial questions like "Is there life in outer space?" Recently, in April 2013, the news that the NASA's Kepler Telescope discovered three new potentially habitable planets (Kepler-22b, -69c, -62e, -62f) in distant solar systems is exciting not only scientists.

Zooming in to the scales of atoms and molecules and investigating their section of the world are providing the information necessary to understand the building blocks of life and the structures and functions of living systems. Mathematicians have become aware very quickly that existing mathematical methods may not be sufficiently adequate for facing the special complexity of biological systems, since the variety and evolvability of biological systems and their intrinsic multi-physics and multiscale structures make them extremely more complex than nonbiological systems. Relevant systems in life sciences are in general coupled networks of subsystems. Biological structures and functions are results of interactions of processes on complex networks. Already Aristotle knew: "The whole is greater than the sum of its parts." The amount and the quality of data in biology and in medicine is increasing due to a rapid improvement in experimental methods and technologies. Processing the available data and deriving the information and concepts necessary to understand and control the systems needs modeling, simulation, and expertise in mathematics, mathematical modeling, and computational sciences, complementing the experimental life sciences. Managing complexity, nonlinearity, high dimensionality, multiple scales, instability, and uncertainty arising in particular in biosystems will be a persistent, long-term challenge. Certainly, advanced computer technology and new algorithms will be important tools to cope with the challenges. However, they have to be complemented with new concepts for modeling the systems, and reducing the models to make them accessible to computation and calibration based on real data. Here multi-core computer systems are offering new perspectives for an integrated approach, including the design of adjusted models and algorithms.

One main aim of the workshop was to identify and discuss proper approaches to overcome some of the arising obstacles and to initiate and support co-operations. The intensive discussions in special sessions had the objective to provide orientation for the future research, based on summaries of the state of the art.

Teamwork could be achieved also in support and organization of the Kepler workshop at the institutional level. The Heidelberg Academy of Sciences and Humanities (HAW) (http://www.haw.uni-heidelberg.de) and the Center for Modelling and Simulation in Biosciences (BIOMS) (http://www.bioms.de) of the University of Heidelberg share with the European Academy of Sciences the aim to promote and support excellent young scientists, interested and engaged in interdisciplinary research and working in teams. Therefore, the Kepler award offers a unique chance for cooperation for the profit of science and the young generation.

On the invitation of HAW, the first part of the workshop took place in the historic building of the Academy, at the foot of the renowned Heidelberg Castle, whereas on the invitation of BIOMS, the second part was hosted in the modern BioQuant Building on the campus of Heidelberg University. The Kepler workshop was integrated in the program of BIOMS, and thus the funding and the local organization could be provided. We are highly grateful for all contributions to this cooperation.

The quality of the scientific contributions, the achieved impulses for future research, and initiated exchange and cooperation are the most rewarding outcomes of the joint activity. The prize winners deserve high appreciation not only for organizing an inspiring workshop, but also for designing and putting together this volume. We hope that the reader will feel some of the exciting atmosphere of the event.

Milano, April 2013                                         Vincenzo Capasso
Heidelberg                                                  Willi Jäger

# Preface

This book is a followup to the scientific workshop "Managing Complexity, Reducing Perplexity" which was held in Heidelberg, May 16–20, 2011, as part of the 2010–11 Kepler Award for European Young Scientists (KEYS), established by the European Academy of Sciences (EURASC). The recipients of the award were Marcello Delitala (Italy) for mathematical sciences, Giulia Ajmone Marsan (France) for social sciences, and Andrea Picco (Germany) for biological sciences. These researchers were chosen from a group of a dozen young European scientists with a Ph.D. in Mathematics, Biology, or Medicine.

"Managing Complexity, Reducing Perplexity" was devoted to an overview of the state of the art in the study of complex systems, with particular focus on the analysis of systems pertaining to living matter. Both senior scientists and young researchers from diverse and prestigious institutions with a deliberately interdisciplinary cut were invited, in order to compare approaches and problems from different disciplines. A common aim of the talks was that of analyzing the complexity of living systems by means of new mathematical paradigms that are more adherent to reality, and which are able to generate both exploratory and predictive models that are capable of achieving a deeper insight into life science phenomena.

The book collects a selection of scientific contributions from the speakers at the meeting.

The interest in complex systems has witnessed a remarkable increase in recent years, due to an increasing awareness that many systems share a common feature, that is "complexity," and that they cannot be successfully modeled by traditional methods used for inert matter systems. According to an opinion that is widely shared in the scientific community, a *Complex System* is any system made up of a large number of heterogeneous interacting entities, whose interactions lead to the emergence of collective behaviors that are not predictable from the individual dynamics. Complex systems are often characterized by nonlinear structures at different representation scales.

When dealing with living systems, it is necessary to face an additional source of complexity: the interacting entities express an individual strategy that modifies classical mechanics laws, and, in some cases, generates proliferative and/or destructive processes. Moreover, the expression of a strategy is heterogeneously distributed over the system. When dealing with living matter, a seminal paper by H. L. Hartwell and co-workers should be recalled:

> The living matter shows substantial differences with respect to the behavior of the inert matter. Although living systems obey the laws of physics and chemistry, the notion of function or purpose differentiates biology from other natural sciences. Organisms exist to reproduce, whereas, outside religious beliefs rocks and stars have no purpose…What really distinguishes biology from physics are survival and reproduction, and the concomitant notion of function.[1]

The interactions between individuals can occur not only through contact, but may be also distributed in space as well as on networks. Collective emerging behaviors, determined by the dynamics of interactions, cannot be described only on the basis of the knowledge of the mechanical dynamics of each element, i.e., the dynamics of a few individuals does not automatically lead to the overall collective dynamics of the whole system.

Thus, complex systems are intrinsically multiscale, and show emerging phenomena at the macroscopic level that express a self-organizing ability, which is the output of the interactions between entities at the microscopic level. Moreover, the emergence is bottom-up, from lower representation to a higher scale, with a feedback loop: the emerging patterns may affect and perturb the lower levels (the so-called immergence: a top-down phenomenon).

Due to this self-organizing ability, feedbacks, and redundancy, in a fast evolutionary framework, complex systems have in many cases a great capacity to adapt to changing landscapes, to cope with environmental changes and pressures, and maintain their structure and stability against the perturbations that occur at various scales.

An increasing number of applications in technology, economics, and social sciences resemble such systems, given their high number of composing elements and the nonlinear connections among them. "Complexity" is one of the main features of a variety of phenomena, from cell biology to fluctuations in economic markets, from the development of communication networks and the Web to traffic flows in highways, to the ecosphere evolution against climate changes, and other generic environmental issues.

Many systems of the physical world are made up of several interconnected components, which may be represented, and, at times, measured, according to different scales of observation. Interactions between different parts of the system may show emerging collective behavior and structures that require specific interpretations for each scale of observation, thus highlighting the new features that arise when passing from one scale to another. Whether you consider the individual entities or their subsets, the simultaneously occurring processes at different temporal and spatial scales characterize the system, so that the laws that govern the behavior of the "whole" are qualitatively different from the laws that govern the individual components.

The investigation on complexity has the objective of understanding what its main properties are. How does the system adapt to evolving conditions? How does

---

[1] H. L. Hartwell, J. J. Hopfield, S. Leibner, and A. W. Murray, From molecular to modular cell biology, *Nature*, 402, c47–c52.

Preface xi

it learn efficiently and how it does optimize its behavior? Are there common rules that govern the behavior of complex systems? The development of a science of complexity cannot be reduced to a single theoretical or technological innovation, but implies a novel scientific approach.

Thus, "managing complexity" means identifying the "complexity" features of a system, modeling its dynamics, highlighting the possible rise of new structures and emerging patterns, investigating their resilience against perturbations, searching for any common features that govern the ways in which this collective behavior occurs. A mathematical approach can provide useful suggestions to help understand the global behavior of a living system by capturing its essential features.

Many mathematical models have been proposed to describe various aspects of complex living systems. There is no universal tool that is more suitable than others: each has its pros and cons, and each aims at highlighting the particular behavior of each particular system at a well-defined level of representation. A research approach should be designed to select the most significant tool to explain the collective behavior, i.e., the tool that contributes the most information for both that particular scale and for the transition from one representation scale to another one.

The description of complex living systems requires challenging mathematical structures and original theories, as well as progress in theoretical methods, in numerical algorithms, and in developing experimental strategies.

Moreover, it is necessary to bring together different kinds of scientific knowledge and different background to tackle this challenging goal: an interdisciplinary approach between scientists from different fields is necessary to define a common protocol that would be able to exchange information, and to design experiments and indicators that can provide information that would enable the validation, and therefore the refinement of already proposed models, to develop qualitative analysis, numerical simulations, and new hypothesis. This is why suitable interactions between groups of researchers from different areas (mathematics, physics, biology, sociology, and economics) are necessary to find new paradigms that can be used to model and investigate a more and more connected, interacting, and globalized world.

The above-mentioned points were common issues in the workshop and will be the key points of this book. The focus is on biological systems; the meeting was in fact devoted to the modeling and simulation in life sciences, focussing on some of the current topics in biology and medicine and the related mathematical methods: several biological systems are characterized by interconnected heterogeneous elements that, together, exhibit some properties, which are often not obvious at first. These systems are demanding for interdisciplinary approaches that are able to combine life sciences and mathematics/physics.

The main topics of the workshop were: complexity in life sciences and in biosystems, regulatory networks, cell motility, multiscale modeling and simulation of cancer, morphogenesis and the formation of biological structures, evolution and adaptation.

These topics have been developed by researchers from various disciplines and different scientific communities (biologists/mathematicians/physicists), who share a common interest in life sciences, with the aim of achieving deeper insight into these biological phenomena and, hopefully, a better understanding, simulation, and control of them.

A key issue that emerged during the discussions was the necessity of more and more direct interaction between Mathematicians/Physicist and Biologists. Indeed, interdisciplinarity was the leading issue of the workshop; the ability to interpret scientific problems from different points of view is evidently more and more important, besides the technical knowledge needed to face them.

Apart from the various talks and discussions, some round table conferences were held that led to some interesting thoughts and outcomes.

The first round table was on specific advice from senior scientists to young ones pertaining to the successful development of scientific research in biomathematics. The results can be briefly synthesized in some memorable sentences that emerged in the discussions:

- *Get wet! Mathematicians perform experiments (F. Bussolino, IRCC, Candiolo, Italy)*
- *Data Driven Modelling together with Model Driven Experiments (V. Capasso, University of Milan, Italy)*
- *Integration. Biologist be your buddy (A. Dell, Imperial College, London)*
- *Modelling, integrating data and concepts of processes (W. Jäger, University of Heidelberg, Germany)*
- *Stay close to the data (V. Quaranta, Vanderbilt University, USA)*
- *Scientific honesty … Do not put all your eggs in one basket (D. Sherrington, University of Oxford, UK)*

The second round table was on which actions are needed by young researchers to support their career development and the need of education for the next generation researchers. Here, it was pointed out that more attention should be paid to graduate education in which the borders of different sectors of sciences are crossed (e.g., Ph.D. programs combining biomedical skills with maths-physics ones), in order to establish a "common protocol" between researchers from different disciplines.

The third round table was on the perspectives of young scientists, in terms of career development and the facility of finding suitable positions. Here, the landscape is heterogeneous, because scientific communities in some countries are still stuck in rigid and classic disciplinary sectors (as, for instance, in Italy), while in other countries (e.g., the UK and the USA) things appear to be different. The suggestion was to try to be truly interdisciplinary, finding stimuli, and looking for new experiences "away from home," if necessary (*Go West, young boy!*).

However, the evident need for a real and continuous interplay between biological sciences and maths-physics emerged from all discussions.

Preface xiii

Another issue that emerged from the discussions was the need for a strong biological approach to reduce the complexity of the system. It is in fact mandatory to develop mathematical tools for each scale that retains the key features of the system. Deeper considerations on this issue have been developed in the last contribution by M. Delitala and T. Hillen.

The book presents 13 contributions dealing with different aspects of complex biological systems.

The book starts with a contribution that frames the problem of dealing with complexity in life sciences and the choice of suitable mathematical methods.

The first contribution by T. Hillen and M. A. Lewis on the Mathematical Ecology of Cancer, highlights other important aspects of dealing with complex systems: the transversality of methods, cross-disciplinarity, and fertilization. Their contribution focuses on the important connections between ecology and cancer modeling, which bring together mathematical oncology and mathematical ecology to initiate cross-fertilization between these fields.

Focusing in more detail on some of the features of complexity, the multiscale nature of these biological systems has been shown in the following three contributions on cancer modeling; the onset and evolution of a tumor is a good example of complex multiscale problem as it is a process that normally spreads over many years and involves a large variety of phenomena that occur at different biological scales.

The chapter by P. Macansantos and V. Quaranta on heterogeneity and growth variability in cell populations focuses on recent advances, both theoretical and experimental, in quantification and modeling of the clonal variability of proliferation rates within cell populations, highlighting work carried out in cancer-related systems.

The contribution by P. Gerlee and S. Nelander is focused on the impact of phenotypic switching in a model of glioblastoma invasion. Simulations of the stochastic model and simulations, obtained by deriving a continuum description of the system, show interesting results on the wave speed of the solutions and suggest a possible way of treating glioblastomas by altering the balance between proliferative and migratory behavior.

The contribution by D. Trucu and M. A. J. Chaplain on Multiscale Analysis and Modelling for Cancer Growth and Development, presents a novel framework that enables a rigorous analysis of processes that occur at three (or more) independent scales (e.g., intracellular, cellular, tissue). Then, a new model is proposed that focuses on the macroscopic dynamics of the distributions of cancer cells and of the surrounding extracellular matrix and its connection with the microscale dynamics of the matrix degrading enzymes, produced at individual cancer cell level.

The need for new mathematical frameworks and tools to deal with some features of the biological phenomena is also evident in the contribution by J. Calvo, J. Soler and M. Verbeni who propose a nonlinear flux-limited model for the transport of morphogens. They introduce flux-limited diffusion as a new tool to obtain mathematical descriptions of biological systems whose fate is controlled by morphogenic proteins.

The biological aspects of multiscale phenomena and the influence of lower scales at macroscopic level is evident in the contribution by A. Dell and F. Sastre on glycosylation: a phenomenon shared by all domains of life. Biological complexity is not linearly related to the number of genes among species: it is well known that the total number of genes in humans is not very different from organisms such as fruit flies and simple plants. The authors point out their attention on a specific phenomenon, the Glycosylation, that occurs after genes have been translated into proteins, and that results in the greatest diversity of the products of gene expression.

The emergence of collective behavior from interactions at a lower lever (including learning, adaptation, and evolutionary dynamics) has been dealt with in detail in the following two contributions.

The chapter by E. Agliari, A. Barra, S. Franz and T. Pentado-Sabetta proposes some thoughts on ontogenesis in B-cell immune networks. It focuses on the antigen-independent maturation of B-cells and, via statistical mechanics tools, studies the emergence of self/non-self-discrimination by mature B lymphocytes and highlights the role of B–B interactions and the learning process at ontogenesis, that develop a stable memory in the network.

In the chapter by M. Delitala and T. Lorenzi, on the mathematical modeling of cancer under target therapeutic actions, the authors focus on emerging behavior in cancer dynamics. Due to the interaction between cells and therapeutical agents, it is shown how competition for resources and therapeutical pressure can lead to the selection of fitting phenotypes and evolutionary behavior, such as drug resistance.

The emergence of patterns and the formation of biological structures is also well represented in the following three contributions.

The contribution by H. Freistühler, J. Fuhrmann and A. Stevens focuses on travelling waves emerging in a diffusive moving filament system. They have derived a model that describes populations of right and left moving filaments with intrinsic velocity, diffusion, and mutual alignment. Analytical investigations and numerical simulations show how interesting patterns are composed of several wave profiles that emerge and the role of different parameters.

The chapter by M. Neuss-Radu on a mathematical model for the migration of hematopoietic stem cells proposes a model, together with a qualitative and computational analysis. The results are compared with experimental results, and possible factors and mechanisms are suggested that can play an important role in emerging behavior to obtain a quantitative description.

The contribution by Jude D. Kong, Sreedhar S. Kumar and Pasquale Palumbo deals with Delay Differential Equation (DDE) models exploited in the specific framework of the glucose-insulin regulatory system, highlighting how those types of models are particularly suited to simulate the pancreatic insulin delivery rate.

The final contributions are related to the different perspectives of management complexity problems in different research fields, and to the different tools that may be employed in the task.

"Physics and Complexity" by D. Sherrington attempts to illustrate how statistical physics has driven the recognition of complex macroscopic behavior as a

# Preface

consequence of the combination of competition and inhomogeneity, and offers new insights and methodologies of wide application that can influence many fields of science.

The last short contribution by M. Delitala and T. Hillen develops some reasonings on the language of Systems Biology and on the need for a multiscale approach to retain some complexity features of the system.

In conclusion, this book has the aim on one hand of offering mathematical tools to deal with the modeling of complex biological systems, and on the other of dealing with a variety of research perspectives. The mathematical methods reported in this book can in fact be developed to study various problems related to the dynamical behavior of complex systems in different fields, from biology to other life sciences. Therefore, applied mathematicians, physicists, and biologists may find interesting hints in this book: to help them in modeling, in developing several analytic problems, in designing new biological experiment, and in exploring new and sometimes unusual perspectives.

This book has been possible thanks to the success of the workshop. Thus, we wish to thank all those who contributed directly or indirectly to the successful organization of the Workshop: the President of the European Academy of Sciences for the initiative of the Award, Vincenzo Capasso, and Willi Jäger for his continuous support, the Direction of BIOMS for the generous financial support, and the local committee of the University of Heidelberg (Willi Jäger, Maria Neuss-Radu, Anna Marciniak-Czochra, and Ina Scheid) for their essential support together with the local Academy of Sciences and Humanities who offered this great opportunity to young researchers and all the speakers and participants. Financial support was also provided by the FIRB project—RBID08PP3J, coordinated by M. Delitala.

Special thanks are due to Prof. T. Hillen, who, in addition to the presentation and the continuous contribution to the activities of the meeting, also collaborated with the concluding contribution of this book.

All information regarding the workshop can be found at the conference website: http://www.eurasc.org/kepler2010.

Turin, November 2012                             Marcello Delitala
Paris                                          Giulia Ajmone Marsan

# Contents

**Mathematical Ecology of Cancer** . . . . . . . . . . . . . . . . . . . . . . . . 1
Thomas Hillen and Mark A. Lewis

**Quantitative Approaches to Heterogeneity and Growth Variability
in Cell Populations** . . . . . . . . . . . . . . . . . . . . . . . . . . . . . . . 15
Priscilla Macansantos and Vito Quaranta

**A Stochastic Model of Glioblastoma Invasion: The Impact
of Phenotypic Switching** . . . . . . . . . . . . . . . . . . . . . . . . . . . . 29
Philip Gerlee and Sven Nelander

**A Hybrid Model for *E. coli* Chemotaxis: From Signaling Pathway
to Pattern Formation**. . . . . . . . . . . . . . . . . . . . . . . . . . . . . . 37
Franziska Matthäus

**Multiscale Analysis and Modelling for Cancer Growth
and Development**. . . . . . . . . . . . . . . . . . . . . . . . . . . . . . . . . 45
Dumitru Trucu and Mark A. J. Chaplain

**A Non-linear Flux-Limited Model for the Transport
of Morphogens** . . . . . . . . . . . . . . . . . . . . . . . . . . . . . . . . . 55
J. Calvo, J. Soler and M. Verbeni

**Glycosylation: A Phenomenon Shared by All Domains of Life** . . . . . . 65
Anne Dell and Federico Sastre

**Some Thoughts on the Ontogenesis in B-Cell Immune Networks**. . . . . 71
Elena Agliari, Adriano Barra, Silvio Franz and Thiago Pentado-Sabetta

**Mathematical Modeling of Cancer Cells Evolution Under
Targeted Chemotherapies** . . . . . . . . . . . . . . . . . . . . . . . . . . . 81
Marcello Delitala and Tommaso Lorenzi

xvii

**Traveling Waves Emerging in a Diffusive Moving Filament System** .................................................. 91
Heinrich Freistühler, Jan Fuhrmann and Angela Stevens

**Homing to the Niche: A Mathematical Model Describing the Chemotactic Migration of Hematopoietic Stem Cells** ........... 101
Maria Neuss-Radu

**DDE Models of the Glucose-Insulin System: A Useful Tool for the Artificial Pancreas** ................................... 109
Jude D. Kong, Sreedhar S. Kumar and Pasquale Palumbo

**Physics and Complexity: An Introduction** .................... 119
David Sherrington

**The Language of Systems Biology** ........................... 131
Marcello Delitala and Thomas Hillen

# Mathematical Ecology of Cancer

**Thomas Hillen and Mark A. Lewis**

> *"The idea of viewing cancer from an ecological perspective has many implications, but fundamentally it means that we cannot just consider cancer as a collection of mutated cells but as part of a complex balance of many interacting cellular and microenvironmental elements"*. (quoted from the website of the Anderson Lab, Moffit Cancer Centre, Tampa Bay, USA.)

**Abstract** It is an emerging understanding that cancer does not describe one disease, or one type of aggressive cell, but, rather, a complicated interaction of many abnormal features and many different cell types, which is situated in a heterogeneous habitat of normal tissue. Hence, as proposed by Gatenby, and Merlo et al., cancer should be seen as an ecosystem; issues such as invasion, competition, predator-prey interaction, mutation, selection, evolution and extinction play an important role in determining outcomes. It is not surprising that many methods from *mathematical ecology* can be adapted to the modeling of cancer. This paper is a statement about the important connections between ecology and cancer modelling. We present a brief overview about relevant similarities and then focus on three aspects; treatment and control, mutations and evolution, and invasion and metastasis. The goal is to spark curiosity and to bring together mathematical oncology and mathematical ecology to initiate cross fertilization between these fields. We believe that, in the long run, ecological methods and models will enable us to move ahead in the design of treatment to fight this devastating disease.

**Keywords** Cancer modelling · Cancer ecology · Microenvironment · Cancer treatment · Mutation and selection · Competition · Immune response

---

T. Hillen (✉) · M. A. Lewis
Centre for Mathematical Biology, University of Alberta, Edmonton, Canada
e-mail: thillen@ualberta.ca

M. A. Lewis
e-mail: mlewis@math.ualberta.ca

M. Delitala and G. Ajmone Marsan (eds.), *Managing Complexity, Reducing Perplexity*,
Springer Proceedings in Mathematics & Statistics 67, DOI: 10.1007/978-3-319-03759-2_1,
© Springer International Publishing Switzerland 2014

# 1 Introduction

The traditional understanding of cancer is based on the view that, through mutations, a very aggressive cell type is created, which grows unlimitedly, is able to evade treatment and, at later stages, invades into other parts of the body (metastasis). All cells of the tumor are considered as basically identical clones. In recent years, however, the picture has changed greatly. It is now well accepted that cancer does not describe one type of aggressive cells, or even one disease, but rather a complicated interaction of many abnormal features (Merlo et al. [46], Hanahan and Weinberg [23, 24] and Gatenby et al. [18, 19]).

A tumor is a result of accumulation of mutations (sometimes 600–1000 mutations), and the tumor mass consists of a heterogeneous mix of cells of different phenotypes. It is these accumulation of mutations which make cancer so dangerous. One mutation might only change a metabolic pathway, but this alone will not suffice for a malignant tumor. As outlined in [24], a full grown invasive tumor can express *cancer stem cells*, which have infinite replicative potential, *progenitor cells* of different abilities, *mesenchymal cells* which result from an endothelial-mesenchymal transition (EMT) and are able to aggressively invade new tissue, *recruited endothelial cells*, which begin to form a vascular network to supply nutrients, *recruited fibroblasts*, which support the physical integrity of the tumor, and *immune cells*, which can be both, tumor-antagonizing and tumor-promoting. All of this resides in a heterogeneous environment of healthy tissue. If such a cancer is challenged by a specific treatment, then only a specific strain of tumor cells will respond to it, and the treatment will select for those cell types that are more resistant to treatment. Hence an immediate consequence of this new understanding is that a single specific treatment is likely to lead to resistance, since only a sub-population is targeted by the treatment. To have any hope of treatment success, a combination therapy should be applied, as is done nowadays in most clinical applications.

Hanahan and Weinberg published a list of six *hallmarks of cancer* in 2000 [23], which has been very highly cited. Just recently [24], in March 2011, they revised their *hallmarks* and adding two *enabling characteristics* and two *emerging hallmarks*. The ten hallmarks, including those of the "next generation" are:

1. sustained proliferative signalling;
2. avoidance of growth suppressors;
3. resistance of cell death;
4. replicative immortality;
5. induction of angiogenesis;
6. invasion and metastasis;
7. genome instability and mutation;
8. deregulation of cellular energetics;
9. tumor promoting inflammation;
10. avoidance of immune destruction.

Hanahan and Weinberg suggest that, to understand tumors, we must look deeper into the microscale processes governing these traits:

> *...tumors are more than insular masses of proliferating cancer cells. Instead they are complex tissues composed of multiple distinct cell types that participate in heterotypic interactions with one another. . . . tumors can no longer be understood simply by enumerating the traits of the cancer cells but instead must encompass the contributions of the "tumor microenvironment" to tumorigenesis.* (page 646 of [24])

This is where dynamical mathematical models play a key role. If hypotheses about the processes at the microscale can be formulated quantitatively, then the dynamics of these processes can form the inputs to a mathematical model, whose analysis then makes predictions about emergent outcomes. The mathematical model thus builds a bridge connecting microscale process dynamics to predicted traits or hallmarks of cancer tumours. A test of the model, and its underlying hypotheses, comes from comparing model predictions for the emerging traits or hallmarks for cancer tumors to actual observations.

## 2 Connecting Ecology to Cancer Modelling

As described above, the process of connecting microscale dynamics to emergent traits is a central endeavour of field of mathematical oncology (see, for example, [1, 30]). However, a similar rubric has also been developed in another subfield of mathematical biology, namely mathematical ecology. Here ecological processes on a small scale are connected to emergent ecosystem properties [42]. The structure of modelling dynamics shares many similarities with the complex interactions between cell types and the environment found in mathematical oncology, although the processes act on organismal rather than cellular scales. However, the area of mathematical ecology was developed earlier than mathematical oncology and so, in some respects, has matured further as a field. The goal of this paper is to draw the connections between mathematical oncology and ecology at the process level, with a view to inspire curiosity and identify areas where *technology transfer* is possible, from one subfield to the other.

The ultimate goal of cancer research is to understand and control cancer growth and to heal the patient. As seen in Hanahan and Weinberg's classification scheme, the process of tumor development, growth and spread is very complex. In addition, inclusion of different treatment modalities, such as surgery, radiation or chemotherapy, makes the whole issue even more complex. Mathematical modelling has helped scientists to navigate through the complicated interactions and to identify basic mechanisms of tumor growth and control. Specifically, models for angiogenesis, for antiangiogenesis, for non-vascular tumor growth and for optimization of chemotherapy or radiation therapy have been used to improve treatment outcomes. Furthermore, mathematical models link the genetic make-up of a cancer to the dynamics of cancer in tissue. It is, however, a long way from a mathematical result to a clinical contribution, and we, as modellers, need to work very hard to convince the medical sciences about the usefullness of mathematical modelling. The ecological community has understood the relevance of modelling already.

Understanding the distribution and abundance of organisms over space and time is the goal of ecology. Mathematical ecology uses quantitative methods to connect the distribution and abundance of organisms to processes such as behavior, competition, food webs, predation, evolution, genetics and environmental fluctuations. Over the past decades, the mathematical modelling of ecosystems has produced some sophisticated theories. For example, there is a vast literature on invasion of foreign species [28], on persistence or permanence of species under stress [3], on bio-control [13] and optimal control [41], on genetics, mutations and selection [34], on competition [60] and predator-prey interactions [27] and many forms of structured population models [7]. Some of these methods have been adapted to the situation of cancer modelling, and we believe that the research on cancer modelling can even further benefit from these methods. Specifically, we see close resemblances between ecology and cancer biology in relation to

(a) **Mutations and Selection**: Genetic instability allows a tumor to adapt to a changing environment, to avoid destruction from the immune system and to evade treatments. In ecology, mutation and selection are the driving principles behind evolution of ecosystems and species.
(b) **Competition**: Cancer cells compete with healthy cells for nutrients. In ecology, many species compete for resources.
(c) **Predator-Prey dynamics**: The immune system can be seen as a predator on the cancer cells. However, the "predator" is not only killing the cancer cells, but might as well promote tumor growth (see [24]).
(d) **Food Chains**: Food chains in ecosystems resemble biochemical pathways and cell metabolism.
(e) **Extinction**: While species extinction is to be avoided in many ecological species, cancer extinction is desired for cancer treatment.
(f) **Age Structure**: Species proliferation naturally depends on the age of the individuals. Similarly, cells are constrained by a cell cycle and they need to transfer through the cell cycle phases ($G_0$, $G_1$, $S$, $G_2$, $M$) before mitosis.
(g) **Periodic Forcing**: Ecosystems underlay day-to-day cycles and seasonal cycles. An important cycle in humans is the circadian rhythm, which has an influence on all cells of the body.
(h) **Cell Movement**: Cancer cells move through a complex heterogeneous tissue network. Similarly animals move through heterogeneous environments. Much work has been done on both, tracking cells (via tagging and microscopy) and tracking animals (via radiocollars and measurement through global positions systems (GPS)).
(i) **Invasions**: Invasions of metastasis is the last step of tumor progression. It is usually responsible for the death of the patient. Invasions of foreign species into native ecosystems is one of the major challenges of modern ecology.
(j) **Fragmentation and patchy spread**: Cancer tumours often appear to be fragmented or patchy. Similarly, population densities are notoriously patchy. Reasons for such patchy distributions, ranging from nonlinear pattern formation to stochastic effects, to environmental heterogeneity, can equally well be applied to

cancer tissues or ecological populations. Furthermore, in ecosystems, the spatial organization of species is an important feature, which enables coexistence of otherwise exclusive species.

(k) **Path generation**: many animal species lay down a network of paths to popular foraging locations. Here we see an analogy to vasculature formation during angiogenesis.

(l) **Control**: Cancer control through treatment resembles ecological control mechanisms such as hunting and harvesting. Also biological control, through parasites, is a possible strategy, which is currently discussed in the context of cancer (e.g. bacterial cancer therapies [15]).

We summarize the relations between cancer and ecology and the type of modelling in the following Table 1. In Fig. 1 we attempt a visual representation of the similarities between these areas.

The resemblance is indeed more than striking, and we can use these relations to our advantage. We should not be shy, but cross borders to benefit from the insights of mathematical ecology. In fact, Merlo et al. [46] write in their abstract on page 924:

> *The tools of evolutionary biology and ecology are providing new insights into neoplastic progression and the clinical control of cancer*

The above list, however, is too wide to be covered in a single short paper. Hence here we will focus on areas that we believe the connections stand out most clearly: *control*, *evolutionary theories* and *cell movement and invasion models*.

## 3 Investigating the Connections

### 3.1 Tumor Control and Treatment

The common therapies against cancer include surgical removal of cancerous tissue, radiation treatment, chemotherapy and hormone therapy. Quite often a combination of these modalities is used (see e.g. [2]). The modelling of the expected treatment success by radiation treatment is an excemplary showcase of cross fertilization between ecology and cancer modelling. The quantity of interest is the *tumor control probability* (TCP). In its simplest form it is given by the *linear quadratic model* [63]

$$TCP = e^{-S(D)}, \qquad S(D) = N_0 e^{-\alpha D - \beta D^2},$$

where $N_0$ denotes the initial number of tumor cells, $S(D)$ denotes the surviving cell number of a treatment with dose $D$, and $\alpha$ and $\beta$ are the radiosensitivity parameters, which depend on the type of tissue and the type of cancer. The TCP describes the probability that a tumor is eradicated by a given treatment. Mathematically, the TCP is the same object as the *extinction probability*, which describes the probability that a certain species of interest (for example an endangered species [48]) goes extinct. Kendal

**Table 1** Similarities between the hallmarks in oncology to ecological mechanisms and typical mathematical models. ODE refers to ordinary differential equations and PDE to partial differential equations. The numbers 1.–10. relate to the hallmarks of Hanahan and Weinberg and the letters (a)–(l) relate to the ecological process as listed above

| Hallmark | Cancer process | Ecological process | Typical model |
| --- | --- | --- | --- |
| 1. Sustained prolif. signalling | Tumor growth | Population growth (d) Food chains | ODE |
| 2. Avoidance of growth suppressors | Tumor growth | Population growth (d) Food chains | ODE |
| 3. Resistance to death | Tumor growth | Population growth (e) Extinctions | ODE stochastic processes |
| 4. Immortality | Tumor growth | Population growth (e) Extinctions | ODE |
| 5. Angiogenesis | Tumor vascularization | (j), (k) Spatial heterogeneity | PDEs stochastic processes |
| 6. Invasion metastasis | Tumor spread | (i), (h) Population spread | PDEs stochastic process |
| 7. Genome instability | Carcinogenesis | (a) Evolution | Integral equations adaptive dynamics |
| 8. Cellular energetics | Competition with healthy tissue | (b) Competition | Dynamical systems |
| 9. Inflammation | Immune response | (c) Predator prey and biological control | Dynamical systems |
| 10. Avoidance of immune destruction | Immune response | (c) Predator prey and biological control | Dynamical systems |
| Additional feature | | | |
| Cell cycle | Cell cycle specific sensitivities | (f) Structured populations | PDEs |
| Treatment | Tumor control | (l) Control of biological species and harvesting | Optimal control problems |
| Periodic forcing | Circardian rythms | (g) Seasonality | Nonautonomous ODEs |

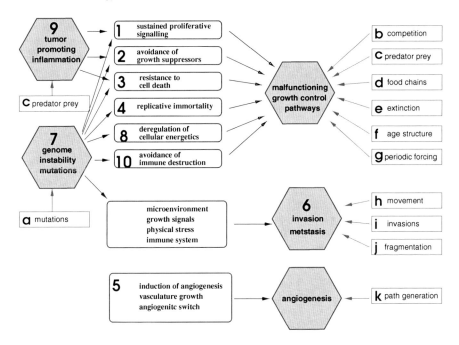

**Fig. 1** Schematic representation of the relations between the cancer hallmarks, ecological processes, and mathematical modelling. The *blue numbers* 1–10 refer to the hallmarks as described by Hanahan and Weinberg [24] and the *green letters* a–k refer to the ecological processes as listed above. The *red hexagons* relate to biological or ecological processes that have been analyzed through mathematical modelling. The *arrows* indicate what kind of information from experiments or observation is used to inform the corresponding models. There are many more feedback loops, from modelling to biology, which we needed to omit due to readability

[37] developed a birth-death framework for the extinction probability, which since has been developed as a more accurate TCP model than the above linear quadratic model. The mathematical framework comes directly from ecological applications, but the interpretations, and some of the details are specific to cancer modelling. This direction of research has blossomed in beautiful theories on brith-death processes and branching processes, which are able to include cell cycle dynamics and differential radiosensitivities depending on the cell cycle state (see [22, 25, 26, 31, 43, 61, 66]). In a recent PhD thesis, Gong [21] included cancer stem cells into the TCP models and she confirmed that it is critical to control the stem cells for treatment to be successful. First studies have shown that the above TCP models are powerful tools in the prediction and planning of radiation treatments ([22, 61]), however, further studies of their qualitative properties and further data analysis is needed.

Ecologists have long assessed the probability of local extirpation of a species of interest using the method of population viability analysis (PVA). This mathematically depicts the birth and death process via a stochastic process with drift, as described by a partial differential equation. Here hitting probabilities and times to extinction can be calculated based on classical diffusion theory [57]. More recently this approach has been modified to address the problem of preventing establishment of a species, rather than preventing extinction of a species. Here the goal is to determining how to prevent introduced exotic species establishing as an invader, with the goal of making them go extinct [9]. This approach shares much with that of controlling cancer.

In a spatial context, ecological modellers have investigated the problem of optimal spatial control of an invader, determining the size and duration of treatment needed to spatially control the spread of an invader as it moves across a landscape [12, 54]. This approach has parallels with the issue of optimal radiation treatment for controlling the spread of a cancer tumour. The optimization of chemotherapy has been the focus of many research groups around the world, for example: Swierniak (Poland); Agur (Israel); Ledzewicz, Schaettler (USA); d'Onofrio, (Italy). A common theme is the occurrence of resistance. We expect that the above mentioned evolutionary theories, can help to better understand the process of tumor resistance.

As outlined above, the understanding of cancer as an ecological system immediately suggests the application of combination therapies including chemotherapy, hormone therapy and radiation. Mathematical optimization of combination therapies has not been carried out in detail but it will be a focus for future studies [2].

## 3.2 Evolution

The important role of mutations and genetic information in carcinogenesis and tumor development is well established. Hanahan and Weinberg [24] include genetic instability as one of the enabling hallmarks, and much of modern cancer research is focussed on gene expressions. However, knowing the genes will not suffice to understand and control cancer. As Gatenby wrote in Nature Reviews 2011 [20] on p. 237:

> A full understanding of cancer biology and therapy through a cataloguing of the cancer genome is unlikely unless it is integrated into an evolutionary and ecological context.

The mathematical modelling of evolution in cancer is in full swing and many methods from ecological modelling are already implemented into cancer modelling. Nagy [49] wrote a review highlighting recent success in the modelling of cancer evolution; Merlo et al. [46] explain cancer as an evolutionary process, and Gatenby [18–20] highlight the interaction between evolution, selection and the tumor microenvironment. Enderling et al. [10] used the genetic makeup of tumor cells to successfully model re-occurence of breast tumors. An emerging focus of interest is the role played by *cancer stem cells* [8, 11, 32].

The mathematical modelling of evolution, selection, mutation, and gene expressions has a long history in ecology [34]. Sophisticated theories include models for adaptive dynamics [6], concepts of evolutionary stable strategies [44], game theoretic approaches [5], and analysis of phylogenetic trees and speciations. Many of these are currently discussed in the context of cancer, in particular to understand development of drug resistance during treatment [35, 38].

The evolutionary theories are strongly connected with all of the other cancer hallmarks. Spatial structure leads to selection pressures on the tumor; spatial niches might arise, where metastasis can form. Related to treatment, each treatment agent forms a selection pressure on the tumor and often resistant tumors develop as a result of treatment.

An important difference between ecology and cancer arises related to the relevant time scales. A generation in a developing tumor can be as short as one cell division cycle. i.e. 1/2 day. Hence selection, adaptation and genetic drifts will show up very quickly. Also, a tumor does not have a long ancestry, which goes back for thousands of generations. Finally, the outcome of a tumor in general, is death and destruction. Hence concepts of survival and fitness need to be understood in the correct context.

## 3.3 Models for Cell Movement and Invasions

The invasion of cancer into healthy tissue is one of the hallmarks of cancer, as described by Hanahan and Weinberg [24]. It is often the last step of a malignant tumor and leads to metastasis and to eventual death of the patient. Recent mathematical modelling has focused on various aspects of tumor invasion. Models are of the form of advection-reaction-diffusion equations and transport equations [55] on the one hand and individual based models (cellular automata [29], Potts model etc., [56]) on the other. The choice of model is largely guided by the available data.

For example, in the lab of Friedl and Wolf [16, 17] in Nijmegen in The Netherlands, individual moving cancer metastasis are visualized by confocal microscopy. Parameters such as mean velocities, mean turning rates and turning angle distributions can be measured. Suitable models on this microscopic scale are individual based models [56], transport equations [30], or stochastic processes [51]. The situation is similar in ecology, where individual movement can be measured through GPS tracking, for example, and also entire populations are observed (e.g. via remote sensing). In ecology a whole range of models is used, from individual based models to population models employing the Fokker Plank equations. Here the challenge arises to combine these approaches and to carefully investigate the transition between scales.

On the other hand, macroscopic data are available that measure the extent of a tumor as a whole. For example MRI imaging of glioma, which show tumor regions and the corresponding edema. For these types of data, we use macroscopic models such as advection-reaction-diffusion models [53]. This process is similar to the biological invasion of an introduced pest species. Here ecologists have a history of characterizing the invasion process by a spreading speed that summarizes the rate at which the population spatially colonizes into the new environment. The approach of using a spreading speed was first pioneered by R. A. Fisher [14] for the spread of an advantageous gene into a new environment, and was later applied in an ecological context by Skellam [59] and many others. It has been modified to include the effects of ecological interactions, such as competition, predator-prey and parasite [58]. More recently authors have shown how long-distance dispersal can dramatically increase spreading speeds [40] and have also assessed the sensitivity of the spreading speed to life history and dispersal parameters [50]. We believe that the metastasis stage in cancer is very similar to the biological invader population with long-distance dispersal, and that the assessment of sensitivity of spreading speeds to local physiological conditions may give new insights into the control of cancer spread.

Related to glioma growth, in recent studies [36, 39, 53, 62], it has been shown that reaction-diffusion models can be used to describe glioma growth in the heterogeneous environment of the brain. The brain is made out of white and grey matter. While the grey matter is mostly homogeneous, the white matter is a fibrous structure. Tumor cells are known to use these fibrous structures to invade new areas. In this context we encounter anisotropic diffusion equations describing different mobility in different directions of the tissue. These models have not yet been analysed in depth and first results show the ability to create unexpected spatial patterns (see e.g. [33, 52]). Interestingly, non-isotropic diffusion models are used to model wolf movement in habitats with seismic lines [33, 45], and again, cross fertilization is imminent.

An important difference between tumors and species arises in relation to the surrounding tissue. A tumor lives in a tissue that consists of healthy cells, blood vessels and structural components of the extracellular matrix (ECM). Hence a growing tumor will exert stress onto the tissue and be exposed to stress from the tissue. The inclusion of these physical properties is challenging and first attempts have been made by Loewengrub et al. [64, 65], Preziosi et al. [47] for tumor growth and by Chaplain and Anderson et al. for angiogenesis [4]. These models take the form of continuum mechanics equations and a whole new skill set is needed to study these models. A careful physics based modelling of tumors in tissue, including the appropriate mechanics, is a necessity and a challenge for modern cancer research.

# 4 Conclusion

Understanding the dynamics of cancer is a major challenge for clinicians. The move towards process-oriented cancer models raises many mathematical and modelling challenges. Indeed, it is often the case that even small changes in model formulation can render a model difficult if not impossible to analyse. Under these circumstances it is natural to draw broadly on the collective knowledge of the research community, embracing results from research problems on similar processes that have arisen in different contexts. Here mathematical ecology has a lot to offer, and the potential impact of moving in this direction of research is imminent.

The goal of this paper is to promote the cross disciplinary exchange of ideas and encourage the reader to assess how methods from one area can be made available to another area. We have made a first step in identifying common mathematical theories and problems and also to identify important differences between ecology and cancer. However, there are many more connections that can be made. Most importantly, we

Mathematical Ecology of Cancer

hope that this work will provide a new approach to harness the powerful mathematical tools used in ecology to further advance the treatment planning of cancer.

**Acknowledgments** We are grateful for discussions with R. Gatenby and P. Hinow, which have motivated us to look deeper into the connection of cancer and ecology. TH acknowledges an NSERC Discovery Grant. MAL acknowledges NSERC Discovery and Accelerator Grants and a Canada Research Chair.

# References

1. T. Alarcon, M.R. Owen, H.M. Byrne, P.K. Maini, Multiscale modelling of tumour growth and therapy: the influence of vessel normalisation on chemotherapy. Comput. Math. Methods Med. **7**(2–3), 85–119 (2006)
2. J.W.N. Bachman, T. Hillen, Mathematical optimization of the combination of radiation and differentiation therapies of cancer. Front Oncol (2013, free online). doi:10.3389/fonc.2013.00052
3. R.S. Cantrell, C. Cosner, V. Hutson, Permanence in ecological systems with spatial heterogeneity. Proc. R. Soc. Edinb. **123A**, 533–559 (1993)
4. M.A. Chaplain, S.R. McDougall, A.R.A. Anderson, Mathematical modeling of tumor-induced angiogenesis. Annu. Rev. Biomed. Eng **8**, 233–257 (2006)
5. T. Day, P. Taylor, Evolutionary dynamics and stability in discrete and continuous games. Evol. Ecol. Res. **5**, 605–613 (2003)
6. U. Dieckmann, Can adaptive dynamics invade. Trends Ecol. Evol. **12**, 128–131 (1997)
7. O. Diekmann, M. Gyllenberg, J.A.J. Metz, H.R. Thieme, On the formulation and analysis of general deterministic structured population models. J. Math. Biol. **36**, 349–388 (1998)
8. D. Dingli, F. Michor, Successful therapy must eradicate cancer stem cells. Stem Cells **24**(12), 2603–2610 (2006)
9. J.M. Drake, D.M. Lodge, Allee effects, propagule pressure and the probability of establishment: risk analysis for biological invasions. Biol. Invasions **8**, 365–375 (2006)
10. H. Enderling, M. Chaplain, A. Anderson, J. Vaidya, A mathematical model of breast cancer development, local treatment and recurrence. J. Theor. Biol. **246**(2), 245–259 (2007)
11. H. Enderling, L. Hlatky, P. Hahnfeldt, Migration rules: tumours are conglomerates of self-metastases. Brit. J. Cancer **100**(12), 1917–1925 (2009)
12. R.S. Epanchin-Niell, A. Hastings, Controlling established invaders: integrating economics and spread dynamics to determine optimal management. Ecol. Lett. **13**, 528–541 (2010)
13. W.F. Fagan, M.A. Lewis, M.G. Neubert, P. van den Driessche, Invasion theory and biological control. Ecol. Lett. **5**, 148–157 (2002)
14. R.A. Fisher, The wave of advance of advantageous genes. Ann. Eugen. Lond. **37**, 355–369 (1937)
15. N.S. Forbes, Engineering the perfect (bacterial) cancer therapy. Nat. Rev. Cancer **10**, 785–794 (2010)
16. P. Friedl, E.B. Bröcker, The biology of cell locomotion within three dimensional extracellular matrix. Cell Motil. Life Sci. **57**, 41–64 (2000)
17. P. Friedl, K. Wolf, Tumour-cell invasion and migration: diversity and escape mechanisms. Nat. Rev. **3**, 362–374 (2003)
18. R.A. Gatenby, J. Brown, T. Vincent, Lessons from applied ecology: cancer control using a evolutionary double bind. Perspect. Cancer Res. **69**(19), 0F1–4 (2009)
19. R.A. Gatenby, R.J. Gillies, A microenvironmental model of carcinogenesis. Nat. Rev. Cancer **8**(1), 56–61 (2008)
20. R.A. Gatenby, R.J. Gillies, Of cancer and cavefish. Nat. Rev. Cancer **11**, 237–238 (2011)

21. J. Gong, Tumor control probability models. Ph.D. thesis, University of Alberta, Canada (2011)
22. J. Gong, M. dos Santos, C. Finlay, T. Hillen, Are more complicated tumor control probability models better? Math. Med. Biol. 19 (2011). doi:10.1093/imammb/dqr023. Accessed 17, Oct 2011
23. D. Hanahan, R. Weinberg, The hallmarks of cancer. Cell **100**(1), 57–70 (2000)
24. D. Hanahan, R.A. Weinberg, Hallmarks of cancer: the next generation. Cell **144**, 646–674 (2011)
25. L.G. Hanin, A stochastic model of tumor response to fractionated radiation: limit theorems and rate of convergence. Math. Biosci. **91**(1), 1–17 (2004)
26. L.G. Hanin, Iterated birth and death process as a model of radiation cell survival. Math. Biosci. **169**(1), 89–107 (2001)
27. M.P. Hassell, *The dynamics of arthropod predator-prey systems* (Princeton University Press, Princeton, 1978)
28. A. Hastings, Models of spatial spread: is the theory compete? Ecology **77**(6), 1675–1679 (1996)
29. H. Hatzikirou, L. Brusch, C. Schaller, M. Simon, A. Deutsch, Prediction of traveling front behavior in a lattice-gas cellular automaton model for tumor invasion. Comput. Math. Appl. **59**, 2326–2339 (2010)
30. T. Hillen, $M^5$ mesoscopic and macroscopic models for mesenchymal motion. J. Math. Biol. **53**(4), 585–616 (2006)
31. T. Hillen, G. de Vries, J. Gong, C. Finlay, From cell population models to tumour control probability: including cell cycle effects. Acta Oncol. **49**, 1315–1323 (2010)
32. T. Hillen, H. Enderling, P. Hahnfeldt, The tumor growth paradox and immune system-mediated selection for cancer stem cells. Bull. Math Biol. **75**(1), 161–184 (2013)
33. T. Hillen, K. Painter, in *Dispersal, Individual Movement and Spatial Ecology: A Mathematical Perspective*, ed. by M. Lewis, P. Maini, S. Petrovskii. Transport and Anisotropic Diffusion Models for Movement in Oriented Habitats (Springer, Heidelberg, 2012), p. 46
34. J. Hofbauer, K. Sigmund, *The Theory of Evolution and Dynamical Systems. London Mathematical Society Student Texts* (Cambridge University Press, Cambridge, 1988)
35. Y. Iwasa, M.A. Nowak, F. Michor, Evolution of resistance during clonal expansion. Genetics **172**, 2557–2566 (2006)
36. A. Jbabdi, E. Mandonnet, H. Duffau, L. Capelle, K.R. Swanson, M. Pelegrini-Issac, R. Guillevin, H. Benali, Simulation of anisotropic growth of low-grade gliomas using diffusion tensor imaging. Magn. Reson. Med. **54**, 616–624 (2005)
37. W.S. Kendal, A closed-form description of tumour control with fractionated radiotherapy and repopulation. Int. J. Radiat. Biol. **73**(2), 207–210 (1998)
38. N.L. Komarova, D. Wodarz, Drug resistance in cancer: principles of emergence and prevention. Proc. Natl. Acad. Sci. USA **102**, 9714–9719 (2005)
39. E. Konukoglu, O. Clatz, P.Y. Bondiau, H. Delignette, N. Ayache, Extrapolation glioma invasion margin in brain magnetic resonance images: suggesting new irradiation margins. Med. Image Anal. **14**, 111–125 (2010)
40. M. Kot, M.A. Lewis, P. van den Driessche, Dispersal data and the spread of invading organisms. Ecology **77**(7), 2027–2042 (1996)
41. S. Lenhart, J.T. Workman, *Optimal Control Applied to Biological Models* (Chapman Hall/CRC Press, London, 2007)
42. S.A. Levin, The problem of pattern and scale in ecology. Ecology **73**(6), 1943–1967 (1992)
43. A. Maler, F. Lutscher, Cell cycle times and the tumor control probability. Math. Med. Biol. **27**(4), 313–342 (2010)
44. J. Maynard-Smith, The theory of games and animal conflict. J. Theor. Biol. **47**, 209–209 (1974)
45. H.W. McKenzie, E.H. Merrill, R.J. Spiteri, M.A. Lewis, Linear features affect predator search time; implications for the functional response. Roy. Soc. Interface Focus **2**, 205–216 (2012)
46. L. Merlo, J. Pepper, B. Reid, C. Maley, Cancer as an evolutionary and ecological process. Nat. Rev. Cancer **6**, 924–935 (2006)
47. F. Mollica, L. Preziosi, and K.R. Rajagopal, (eds.), *Modelling of Biological Material* (Birkhauser, New York, 2007)

48. W.F. Morris, D.F. Doak, *Quantitative Conservation Biology: Theory and Practice of Population Viability Analysis* (Sinauer Associates Inc., Sunderland, 2002)
49. J.D. Nagy, The ecology and evolutionary biology of cancer: a review of mathematical models for necrosis and tumor cell diversity. Math. Biosci. Eng. **2**(2), 381–418 (2005)
50. M.G. Neubert, H. Caswell, Demography and dispersal: calculation and sensitivity analysis of invasion speed for stage-structured populations. Ecology **81**, 1613–1628 (2000)
51. H.G. Othmer, S.R. Dunbar, W. Alt, Models of dispersal in biological systems. J. Math. Biol. **26**, 263–298 (1988)
52. K.J. Painter, Modelling migration strategies in the extracellular matrix. J. Math. Biol. **58**, 511–543 (2009)
53. K.J. Painter, T. Hillen, Mathematical modelling of glioma growth: the use of diffusion tensor imaging DTI data to predict the anisotropic pathways of cancer invasion (2012) (submitted)
54. A.B. Potapov, M.A. Lewis, D.C. Finnoff, Optimal control of biological invasions in lake networkds. Nat. Resour. Model. **20**, 351–380 (2007)
55. L. Preziosi (ed.), *Cancer Modelling and Simulation* (Chapman Hall/CRC Press, Boca Raton, 2003)
56. K.A. Rejniak, A.R.A. Anderson, Hybrid models of tumor growth. WIREs Syst. Biol. Med. **3**, 115–125 (2011)
57. E. Renshaw, *Modelling Biological Populations in Space and Time* (Cambridge University Press, Cambridge, 1991)
58. N. Shigesada, K. Kawasaki, *Biological Invasions: Theory and Practice* (Oxford University Press, Oxford, 1997)
59. J.G. Skellam, Random dispersal in theoretical populations. Biometrika **38**, 196–218 (1951)
60. H. Smith, *The Theory of the Chemostat* (Cambridge University Press, Cambridge, 1995)
61. N.A. Stavreva, P.V. Stavrev, B. Warkentin, B.G. Fallone, Investigating the effect of cell repopulation on the tumor response to fractionated external radiotherapy. Med. Phys. **30**(5), 735–742 (2003)
62. K.R. Swanson, C. Bridge, J.D. Murray, E.C. Jr Alvord, Virtual and real brain tumors: using mathematical modeling to quantify glioma growth and invasion. J. Neurol. Sci. **216**, 110 (2003)
63. T.E. Weldon, *Mathematical Models in Cancer Research* (Adam Hilger, Philadelphia, 1988)
64. S.M. Wise, J.A. Lowengrub, H.B. Frieboes, V. Cristini, Three-dimensional multispecies nonlinear tumor growth—I. J. Theor. Biol. **253**, 524–543 (2008)
65. H. Youssefpour, X. Li, A.D. Lander, J.S. Lowengrub, Multispecies model of cell lineages and feedback control in solid tumors. J. Theor. Biol. **304**, 39–59 (2012)
66. M. Zaider, G.N. Minerbo, Tumor control probability: a formulation applicable to any temporal protocol of dose delivery. Phys. Med. Biol. **45**, 279–293 (2000)

# Quantitative Approaches to Heterogeneity and Growth Variability in Cell Populations

**Priscilla Macansantos and Vito Quaranta**

**Abstract** Clonal heterogeneity in cell populations with respect to properties such as growth rate, motility, metabolism or signaling, has been observed for some time. Unraveling the dynamics and the mechanisms giving rise to such variability has been the goal of recent work, largely aided by quantitative/ mathematical tools. Quantitative evaluation of cell-to-cell variability (heterogeneity) poses technical challenges that only recently are being overcome. Clearly, a mathematical theory of cellular heterogeneity could have fundamental implications. For instance, a theory of cell population growth variability, coupled with experimental measurements, may in the long term be crucial for an in-depth understanding of physiological processes such as stem cell expansion, embryonic development, tissue regeneration, or of pathological ones (e.g., cancer, fibrosis, tissue degeneration). We focus on recent advances, both theoretical and experimental, in quantification and modeling of the clonal variability of proliferation rates within cell populations. Our aim is to highlight a few stimulating examples from this fledgling and exciting field, in order to frame the issue and point to challenges and opportunities that lie ahead. Furthermore, we emphasize work carried out in cancer-related systems.

**Keywords** Heterogeneity · Cell-to-cell variability · Systems biology · Cancer · Mathematical modeling of heterogeneity

---

P. Macansantos (✉)
Department of Mathematics and Computer Science, University of the Philippines Baguio, Baguio, Phillipines
e-mail: pmacansantos@yahoo.com

V. Quaranta
NCI Center for Cancer Systems Biology and Department of Cancer Biology, Vanderbilt University School of Medicine, Nashville, TN 37232, USA
e-mail: vito.quaranta@Vanderbilt.Edu

M. Delitala and G. Ajmone Marsan (eds.), *Managing Complexity, Reducing Perplexity*,
Springer Proceedings in Mathematics & Statistics 67, DOI: 10.1007/978-3-319-03759-2_2,
© Springer International Publishing Switzerland 2014

# 1 Heterogeneity and Growth Variability

Clonal heterogeneity in cell populations with respect to properties such as growth (Fig. 1), motility (Fig. 2), metabolism or signalling (Fig. 3), has been observed for some time (see, e.g., Altschuler [1], Quaranta [12] and references therein). However, quantitative evaluation of this cell-to-cell variability (heterogeneity) poses technical challenges that only recently are being overcome [1, 12]. Furthermore, mathematical foundations for interpreting these quantitative experimental observations of heterogeneity are in need of development. Far from being exclusively academic, a mathematical theory of cellular heterogeneity could have fundamental implications, similar to a theory on population biology or ecology [8]. For instance, a theory of cell population growth variability, coupled to experimental measurements, may in the long term be crucial for an in-depth understanding of physiological processes such as stem cell expansion, embryonic development, tissue regeneration, or of pathological ones (e.g., cancer, fibrosis, tissue degeneration).

Here, we focus on recent advances, both theoretical and experimental, in quantification and modeling of the clonal variability of proliferation rates within cell populations. Our aim is to highlight a few stimulating examples from this fledgling and exciting field, in order to frame the issue and point to challenges and opportunities that lie ahead. Furthermore, we emphasize work carried out in cancer-related systems. As our aim is not an exhaustive review, we apologize in advance for inevitable omissions.

Variability of growth rates, among other indicators of heterogeneity in growth kinetics of individual tumours, has long been detected, but precision in quantification may have been made possible only in the past few years by methods developed by, among others, Quaranta and his group (see [12, 19]). For instance, a team from Verona, Italy, quantified growth variability of tumour cell clones from a human leukaemia cell line, by cloning Molt3 cells, and measuring the growth of 201 clonal populations by microplate spectrophotometry. Growth rate of each clonal population was estimated by fitting data with the logistic equation for population growth [18]. Their results indicated that growth rates vary between clones. Six clones with growth rates above or below the mean growth rate of the parent population were further cloned, and the growth rates of their offspring were measured. Researchers noted that distribution of subclone growth rates did not significantly differ from that of the parent population, supporting the conjecture that growth variability has an epigenetic origin [18]. Such variability in growth rates may be amenable to further quantitative analysis of population dynamics with analytic tools developed in Tyson et al. [19].

In the paper "Characterizing heterogeneous cellular responses to perturbations" [14], Slack et al. approached the challenge of heterogeneity with a mathematically-appealing assumption that cell populations may be described as mixtures of a limited number of phenotypically distinct subpopulations. Methods for characterizing spatial heterogeneity observed within cell populations are developed, starting from the extraction of phenotypic measurements of the activation and colocalization patterns of cellular readouts from large numbers of cells in diverse conditions. Phenotypic

Quantitative Approaches to Heterogeneity and Growth Variability in Cell Populations 17

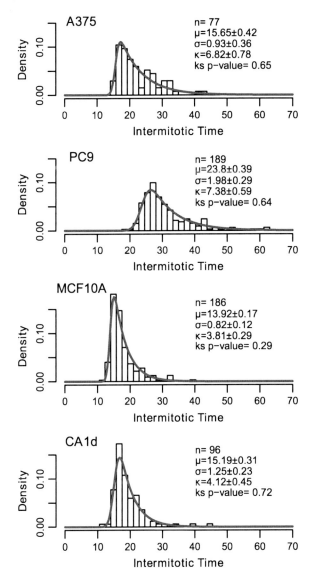

**Fig. 1** Cell-to-cell variability of intermitotic times within human cultured cell lines. Note that the heterogeneity of intermitotic times within seemingly homogeneous isogenic cell lines (populations) is quite broad, and distributed in non-Gaussian fashion. Intermitotic time encompasses hours from the end of one cell division to the start of the successive one. Single cells were tracked by automated confocal microscopy collecting images at regular intervals by automated microscopy as described [19]. Intermitotic times were calculated as described [19] and fitted to an exponentially modified Gaussian (EMG) distribution http://en.wikipedia.org/wiki/Exponentially_modified_Gaussian_distribution. Human cultured cell lines are as follows: A375, melanoma; PC9, non-small cell lung carcinoma; MCF10A, immortalized non-tumorigenic breast epithelium; CA1D, H-Ras transformed MCF10A. $n$ = number of cells tracked; $\mu$, $\sigma$ and $\kappa$ are parameters for the EMG distribution; ks $p$-value was calculated by the Kolmogorov-Smirnoff statistic test.

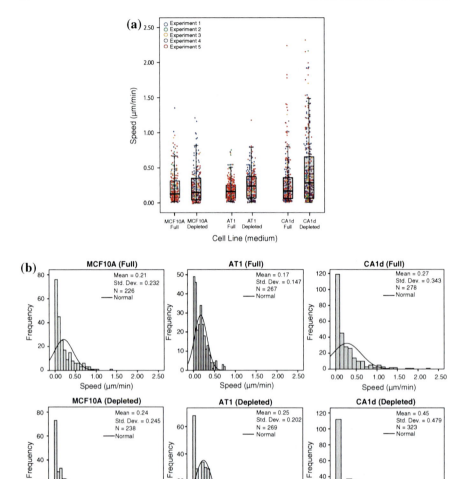

**Fig. 2** Cell-to-cell speed variation within mammary gland human cell lines. Spontaneous, non-directed motility was tracked in over 1,500 individual cells from one immortalized (MCF10A) and two transformed MCF10A-derived (AT1 and CA1d) breast epithelial cell lines. Cell-to-cell variability of motility was evaluated with respect to speed under two culture conditions, full-supplement or serum/EGF-depleted media, respectively. **a** Box-and-whisker plot of individual cell speed (color-coded by individual experiment). **b** Population histogram of frequency (the number of cells) and the normal (Gaussian) fit for each set of data (based around the average). Shapiro-Wilks W tests confirmed that distributions are non-normal and positively skewed (more cells are likely to move at lower speeds) with long tails (at higher speeds).

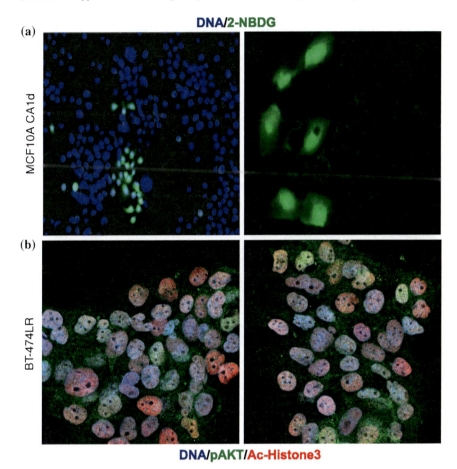

**Fig. 3** Single cell variability in metabolic and signaling activity. **a** Single-cell measurements of glucose uptake using 2- deoxy- 2- [(7- nitro- 2, 1, 3- benzoxadiazol- 4- yl)amino]- D- glucose (2-NBDG). Fluorescent representative images of CA1d (right, higher magnification) after 10 min incubation with 300 $\mu$M of 2-NBDG as described in [7]. The variability in subcellular distribution of the probe was apparent in CA1d cells (right panel). **b** Staining patters of BT-474 lapatinib resistant cell line reveals variability (heterogeneity). BT-475LR cell lines were plated overnight and treated with 1 $\mu$M lapatinib for 1 h at 37 C. Cells were fixed and stained with fluorescent probes (DNA/pAKT(pS473)/Ac-Histone3) and *imaged* with a Zeiss confocal microscope (LSM 510).

stereotypes are identified within the total population, and probabilities assigned to cells belonging to subpopulations modeled on these stereotypes. Each population or condition may then be characterized by a probability vector—its subpopulation profile—estimating the number of cells in each subpopulation. Responses of heterogeneous cellular populations to perturbations (e.g., anti-cancer drugs) are summarized as probabilistic redistributions of these mixtures. In the study by Slack et al., this computational method was applied to heterogeneous responses of cancer cells to a panel of drugs. The finding is that cells treated with drugs of similar mechanism exhibited the same pattern of heterogeneity redistribution.

In subsequent work from this group, Singh et al. [13] employed the same computational framework to investigate whether patterns of basal signaling heterogeneity in untreated cell populations could distinguish cellular populations with different drug sensitivities. As in the earlier study, cellular heterogeneity in populations was modeled as a mixture of stereotyped signaling states. Interestingly, the researchers found that patterns of heterogeneity could be used to separate the most sensitive and most resistant populations to the drug paclitaxel within a set of H460 lung cancer clones and within the NCI-60 panel of cancer cell lines, but not for a set of less heterogeneous immortalized noncancer HBEC (human bronchial epithelial cell) clones. Stockholm et al. [17] used both computer simulation and experimental analysis to address the issue of the origin of phenotypic differentiation in clonal populations. Two models—referred to as the "extrinsic" and "intrinsic" models—explaining the generation of diverse cell types in a homogeneous population, were tested using simple multi-agent computer modeling. The approach takes each cell as an autonomous "agent", and following defined rules governing the action of individual agents, the behavior of the system emerges as an outcome of the agents' collective action.

As the term suggests, the "extrinsic" model attributes the occurrence of a phenotypic switch to extrinsic factors. Identical cells may become different because they encounter different local environments that induce alternative adaptive responses. Changing its phenotype, the cell contributes to changes in the local environment, inducing responses in surrounding cells, and ultimately influencing the dynamics of the cell population. The second model assumes that the phenotype switch is intrinsic to the cells. Phenotypic changes could occur even in a homogenous environment and may result from asymmetric segregation of intrinsic fate determinants during cell division that lead to the change in gene expression patterns, [17].

The Stockholm study cites an experiment where two subpopulations appear spontaneously in C2C12 mouse myogenic cells—the main population (MP), and a side population (SP). The two cell types are phenotypically distinct, and researchers take off from the lab experiment to perform agent-based modeling computer simulation on two cell types subject to two sets of hypotheses (the extrinsic and intrinsic models). The models are built on a limited number of simplified assumptions about how individual cells migrate, interact with each other, divide and die. The agent-based model assumes that each cell divides at each iteration step but survival of daughter cells depends on local cell density. In the intrinsic model, the phenotypic switch occurs under the assumption of cell autonomy, with the environment playing no ostensible role in the switching; rather, switching from one cell type to the other occurs at fixed probabilities. In the extrinsic model, local cell density determines phenotypic switching, hence local density is surrogate for the complex of factors affecting cell survival, such as gradient of nutrients, oxygen, secreted factors, etc, and cell types represent two forms of adaptation to high and low density environments. The extrinsic and intrinsic hypotheses were implemented by varying the parameters (assuming cell migration velocities within experimentally guided limits of values). Simulations for the intrinsic model result in the two cell types being distributed randomly both during growth and equilibrium, suggesting that the randomness of cell type spatial distribution is characteristic of the intrinsic model. On the other hand, the spatial distribution of cells resulting from simulations of the extrinsic model is different from that in the intrinsic model, with cluster formation as an observed feature. Moreover, this feature is robust in the range of parameter values considered.

Both intrinsic and extrinsic models generate in the simulations heterogeneous cell populations with a stable proportion of the two cell types. Experimental verification of model predictions, using the C2C12 myogenic cell line, indicated that neither one of the models can fully account for the spatial distribution of the cell types at equilibrium, as some clustering of the rare SP cell was observed in low density regions, while distribution in high density regions was generally uniform. A hybrid model combining both intrinsic and extrinsic hypotheses was in better agreement with the clustering behavior of the rare SP cells. In the end, it is not solely the local environment, nor, on the other hand, merely a cell-autonomous propensity for differentiation that activates the phenotype switch. Rather, it may be a combination of the two.

A similar "agent" model framework is utilized in mathematical models of cancer invasion, with emphasis on tumor microenvironment, compared in [11]. In that review, three independent computational models for cancer progression are discussed, all pointing to an essential role of the tumor microenvironment (mE) *"in eliciting invasive patterns of tumor growth and enabling dominance of aggressive cell phenotypes."* Both the evolutionary hybrid cellular automata (EHCA) and the Hybrid Discrete Continuum (HDC) models treat cells as points on a lattice. In the case of the EHCA, the grid itself represents the mE, and the only variable on the grid, apart from cells, is the concentration of oxygen, with a partial differential equation controlling the oxygen dynamics in space and time. In the HDC model, the mE consists of a two-dimensional lattice of extracellular matrix upon which oxygen diffuses and is produced/consumed, and matrix degrading proteases are produced/used. The HDC model has the mE variables controlled by reaction-diffusion equations with tumor cells occupying discrete lattice points. Notably, a key feature of the HDC model is that the tumor cell population is heterogeneous, each cell phenotype being defined from a pool of 100 pre-defined phenotypes within a biologically relevant range of cell-specific traits. Mutation is incorporated into the model by assigning to cells a small probability of changing some traits at cell division. If a change occurs, the cell is randomly assigned a new phenotype from the pool of about 100. Taken together with a third model—the Immersed Boundary method (IBCell)—the models describe the process of cancer invasion on multiple scales: The EHCA at the molecular (gene expression) scale, the IBCell at the cell scale, the HDC at the tissue scale. Though not highlighted, heterogeneity is an issue addressed in the models, with the microenvironment driving cancer progression in a major way, and on multiple scales. From representative simulations of the models (see [11] for details), analysis of the effect of mE variables on tumor growth point to *"competitive adaptation to mE conditions as a determining factor for invasion: both invasive tumor morphology ("fingering") and evolution of dominant aggressive clonal phenotypes appear to occur by a process of progressive cell adaptation to mE's that support sustained competition between distinct cancer cell phenotypes."*

In their 2011 paper [6] on models of heterogeneous cell populations, Hasenauer et al. discuss a framework for modeling genetic and epigenetic differences among cells. With the approach to intracellular biochemical reaction networks modeled by systems of differential equations (which may characterize metabolic networks and signal transduction pathways), heterogeneity in populations is accounted for by differences in parameter values and initial conditions. Using population snapshot data, a Bayesian approach is used to infer parameter density of the model describing single cell dynamics. Using maximum likelihood methods, single cell measurement data is processed for parameter density estimation; the proposed framework includes a noise model, as well as methods for determining uncertainty of the parameter density. For computational tractability, the population model is converted into a density-based model, where the variables are not states of single cells but density of the output (see [6] for details). Towards verifying efficacy of the proposed modeling framework, the model of TNF (tumor necrosis factor) signaling pathway was studied under a hypothetical experimental set-up with artificial data involving a cell population responding to the TNF stimulus. The model, introduced in [3], is based on known inhibitory and activating interactions among key signaling proteins of the TNF pathway. Cellular response to the TNF stimulus has been observed to be highly heterogeneous within a clonal population. Heterogeneity at the cell level is modeled by differences in two parameter values, one quantifying the inhibitory effect of NF-êB via the C3a inhibitor XIAP onto the C3 activity, and the other the activation of I-kB via NF-kB. The authors conclude that the method yields good estimation results.

In the abovementioned framework, the assumption was that network structure was identical in all cells and spatial effects and stochasticity of the biochemical reactions are negligible. Moreover, the mechanisms for cell-to-cell interactions typically characterized by differential equations, are reasonably well-understood and formulated, from actual experiment.

In an effort to uncover sources of cell-to-cell variation, Colman-Lerner et al. [4] looked into cell-to-cell variability of a prototypical eukaryotic cell fate decision system, the mating pheromone response pathway in yeast. Cell-to-cell variation was quantified by the output in the cell-fate decision system—the pheromone response pathway in the yeast *Saccharomyces cerevisiae*. The fate decision to switch from the normal vegetative growth to mating events including gene transcription, cell cycle arrest, etc. is induced by the alpha-factor, a pheromone secreted by cells of the mating type. Pheromone-induced expression of fluorescent protein reporter genes was used as a readout. To dis-aggregate differences due to the operation of the signal transduction pathway from cell-to-cell differences in gene expression from the reporters, yeast strains containing genes for the yellow and cyan fluorescent protein were generated. The analytical framework used considered the alpha-factor response pathway and the reporter gene expression mechanism to measure its activity as a single system, with two connected subsystems—pathway and expression. In each of the two subsystems, two sources of variation are considered—stochastic fluctuations and cell-to-cell differences in "capacity", depending on number, localization and activity of proteins that transmit the signal (pathway capacity) or express genes into proteins (expression capacity). About half of the observed variation was attributed to pre-existing differences in cell cycle position at the time of pathway induction, while another large component of the variation in system output is due to differences in cell capacity to express proteins from genes. Very little variation is due to noise in gene expression. Although the study did not specifically refer to molecular mechanisms underlying cell-to-cell variation, it does provide a basis for further investigation into these mechanisms, including, as mentioned elsewhere, network architecture.

Heterogeneous cell populations have been the subject of mathematical modeling since about the 1960s, with the cell population balance (CPB) approach by Frederickson and a few others (see [16] for references). The models use partial integro-differential equations for the dynamics of the distribution of the physiological state of cells and ordinary integro- differential equations to describe substrate availability. For CPB models, heterogeneity arises from physiological functions leading to different growth and division rates of the cells, as well as for unequal partitioning effects. When the physiological state vector (whose components include intracellular content, morphometric characteristics like size) has two or more components, the approach leads to multidimensional models that are highly unwieldy computationally. Stamatakis notes that CPB models cannot account for the inherent stochasticity of chemical reactions occurring in cellular control volumes or stochastic DNA-duplication. To account for this stochasticity, refinements were considered by Gillespie and others (see [16] for references) using the chemical master equation. A relatively recent approach, referred to as the Langevin approach, uses stochastic differential equations in modeling stochasticity in intracellular reactions. In recent work Stamakis and Zygourakis (2010) [16] propose a mathematical framework to account for all the various sources of cell population heterogeneity, namely growth rate variability, stochasticity in DNA duplication and cell division, and stochastic reaction occurrences for the genetic network, through the cell population master equation (CPME) that governs the temporal dynamics of the probability of finding the cell population at a specific state, together with a Monte Carlo algorithm that enables simulation of exact stochastic paths of this master equation. Employing the population balance framework, each cell is described by a state vector containing information about its chemical content and morphometric characteristics such as length, etc (Stamatakis uses volume only). The state of the overall population is given by a vector $\mathbf{w}$, which reflects the number $v$ of individual cells and the state of each vector. The master equation is derived as a probability balance describing the evolution of a probability distribution for the cell population, using submodels of probability inflows and probability outflows accounting for chemical reactions, DNA duplication, cell growth (here using exponential growth), a propensity function (for cells to divide).

In an earlier study, Mantzaris [9] also looked into models of cell population heterogeneity, incorporating into a prior deterministic single-cell model, two extra parameters (one, a rate of operator fluctuations) to quantify two main sources of stochasticity at the single cell level for the reaction network, namely small number of molecules and slow operator fluctuations. Starting from a deterministic cell population balance model (DCPB), Mantzaris used stochastic differential equations to refine the CPB model (to account for extrinsic and intrinsic sources of population heterogeneity—respectively, the unequal partitioning between daughter cells of intracellular components on division, and random fluctuations in reaction rates regulated by a small number of regulatory molecules) through the Stochastic Variable Number Monte Carlo method/model. Simulations on a genetic network with positive feedback revealed differences arising from different sources of stochasticity on regions of the parameter space where the system is bistable.

Although much of the modeling of heterogeneity has not specifically investigated implications on cancer treatment, a 2012 study (see [10]) looks into cell-cycle heterogeneity and its effects on solid tumor response to chemotherapy. In their paper, Powathil et al. raise the difficulty of treating cancer with chemotherapeutic drugs due to the development of cell-cycle mediated drug resistance. Elsewhere (see references in [10]) it has been suggested that this may be due to the presence of functionally heterogeneous cells and can be addressed to some extent by using combinations of chemotherapy drugs that target different phases of the cell-cycle kinetics. Hence, it is important to study and analyze the underlying heterogeneity within a cell and within a solid tumour due to the presence of the unfavourable microenvironment and the cellcycle position. A hybrid multi-scale cellular automaton model is used to simulate the spatio-temporal dynamics at the cell level, incorporating feedbacks between these cell level dynamics and molecular variations of intercellular signalling and macroscopic behaviour of tissue oxygen dynamics. Each cell has its own cell-cycle dynamics and this is incorporated into the CA model for cellular proliferation using a set of ordinary differential equations, from an early model by Tyson and Novak [20]. Chemical processes within the cell are quantified using concentration of key chemical components, considered as functions of time, and a 6-variable system of differential equations describe the processes of production, destruction and interactions. These kinetic relations are then used to explain transitions between two steady states—the G1 and the S-G2-M state, assumed to be controlled by cell mass. With cells located spatially in the dynamic microenvironment, depending on variations in oxygen concentration and with drug distribution dynamics in the growing tumor also affecting the state of individual cells, partial differential equations (for oxygen, a reaction diffusion equation) model changes in oxygen and drug concentration. In simulating the model, parameters were chosen based on earlier work (mainly from Tyson and Novak); notably, to account for the "natural" variability between cell growth rates, and to have a non-synchronous cell population, a multiple of the value from a probability density function with uniform distribution between -1 and 1 is added to an identified value for growth rate, effectively incorporating cell cycle heterogeneity. Computational simulations were run first on cell-cycle and oxygen tumor growth, assuming zero drug concentration, and subsequently on tumours treated with cell-cycle specific drugs. The results revealed that cytotoxic effect of combination therapy depends on timing of drug delivery, time-delay between doses of chemotherapeutic drugs, and cell-cycle heterogeneity. Not surprisingly, drug effectiveness also depends on distribution of tumor cell mass as it affects the tumor microenvironment and drug distribution. The current direction towards patient specific optimal treatment strategies seems to be supported by the model simulations. It is worth noting that non-synchronous cell population can be parameterized from experimental data [5] due to recent automated microscopy advances, making it possible to validate models such as the one described by Powathil et al. [10].

In a recent review by Bendall and Nolan [2], the authors assert that *"stem cell hierarchies, transcription start sites, cell signaling pathways (and more) all function against a backdrop that assumes that carefully orchestrated single-cell stochastics, in concert with mass action, is what determines outcome."* Since all kinds of heterogeneity may drive treatment decisions, it is crucial to develop better technologies to study heterogeneity in single-cells. Notably, the statement is made that recent research indicate that the biology of single cells *"is rarely deterministic."* Snijder and Pelkmans [15] take the view instead that *"a large part of phenotypic cell-to-cell variability is the result of deterministic regulatory processes."* Although not necessarily in conflict, these seemingly opposing views point to the necessity to further investigate various and diverse aspects and mechanisms driving phenotypic heterogeneity in cells and cell populations. As Snijder points out, population context has been shown to contribute in major ways to cellular behavior, including sporulation, genetic competence and motility, giving rise to adaptation in gene transcription, protein translation, cellular growth, rate of proliferation, sensitivity to apoptosis, metabolic activity, cell shape and/or cell polarization. These adaptations cause cells themselves to alter population context, eventually determining single-cell distribution of phenotype properties in a population. Such complex feedback/ regulatory mechanisms may involve many entities and interactions, in the absence of a full understanding of which, a stochastic distribution may somewhat account for the variability [21].

## 2 Conclusions

What emerges from the models so far developed is that apparently "stochastic/variable behavior" in single cells and populations can be reasonably quantified, if not fully understood. In many of the above-mentioned mathematical models for population heterogeneity, the key to characterization of population behavior is a fairly holistic understanding of the key "players" (cells), their environment, and reactions and feedback mechanisms among components. Integration of these theoretical and quantitative tools will be paramount for distinguishing between relevant and noisy heterogeneity [1]. While this field of investigation is still in its infancy, it is not difficult to imagine the impact it will have on our understanding of cellular response to perturbations, including drugs.

**Acknowledgments** We would like to thank the contributors of data included: Fig. 1—Darren R. Tyson (Center for Cancer Systems Biology Center and Department of Cancer Biology, Vanderbilt School of Medicine) and Peter L. Frick (Center for Cancer Systems Biology Center and Chemical and Physical Biology Program, Vanderbilt School of Medicine); Fig. 2—Mark P. Harris (NextGxDx, Nashville, TN) and Shawn P. Garbett (Center for Cancer System Biology and Department of Cancer Biology, Vanderbilt School of Medicine); and Fig. 3—Mohamed Hassanein (Pulmonary Medicine,

Vanderbilt School of Medicine) and Hironobu Yamashita (Department of Gynecology and Obstetrics, Tohoku university, Sendai, Japan). We would also like to thank all the members of the Quaranta laboratory for useful discussions and Lourdes Estrada for discussing key aspects of the manuscript. Furthermore, we would like to acknowledge the support from the Integrative Cancer Biology Program (ICBP) U54CA113007 (V.Q.) and the Philippines' Commission on Higher Education for funding Dr. Priscilla Macansantos' visit to Vanderbilt School of Medicine.

# References

1. S.J. Altschuler, L.F. Wu, Cellular heterogeneity: do differences make a difference? *Cell* **141** (4), 63–559 (2010)
2. S.C. Bendall, G.P. Nolan, From single cells to deep phenotypes in cancer. Nat. Biotechnol. **30**(7), 47–639 (2012)
3. M. Chaves, T. Eissing, F. Allgöwer, Bistable biological systems: a characterization through local compact input-to-state stability. *IEEE Trans. Autom. Control.* (8ADNov.) **13**(53), 87–100 (2008)
4. A. Colman-Lerner, A. Gordon, E. Serra, T. Chin, O. Resnekov, D. Endy, et al. Regulated cell-to-cell variation in a cell-fate decision system. *Nature 2005th ed.* **437**(7059), 699–706 (2005)
5. P. Gabriel, S.P. Garbett, V. Quaranta, D.R. Tyson, G.F. Webb, The contribution of age structure to cell population responses to targeted therapeutics. J. Theor. Biol. **311**, 19–27 (2012)
6. J. Hasenauer, S. Waldherr, M. Doszczak, N. Radde, P. Scheurich, F. Allgöwer, Identification of models of heterogeneous cell populations from population snapshot data. BMC Bioinform. **12**, 125 (2011)
7. M. Hassanein, B. Weidow, E. Koehler, N. Bakane, S. Garbett, Y. Shyr et al. Development of high-throughput quantitative assays for glucose uptake in cancer cell lines. *Mol. Imaging. Biol.* **13**(5), 52–840 (2011)
8. C.J. Krebs, in The problem of abundance: Populations, ed. by B Cummings. *Ecology: The Experimental Analysis of Distribution And Abundance*, 6th edn. Benjamin/Cummings, Menlo Park, 2008), p. 111
9. N.V. Mantzaris, From single-cell genetic architecture to cell population dynamics: quantitatively decomposing the effects of different population heterogeneity sources for a genetic network with positive feedback architecture. Biophys. J. **92**(12), 88–4271 (2007)
10. G.G. Powathil, K.E. Gordon, L.A. Hill, M.A.J. Chaplain, Modelling the effects of cellcycle heterogeneity on the response of a solid tumour to chemotherapy: biological insights from a hybrid multiscale cellular automaton model. J. Theor. Biol. **308**, 1–19 (2012)
11. V. Quaranta, K.A. Rejniak, P. Gerlee, Anderson AR. Invasion emerges from cancer cell adaptation to competitive microenvironments: quantitative predictions from multiscale mathematical models. *Semin. Cancer Biol. 2008th ed.* **18**(5), 338–48 (2008)
12. V. Quaranta, D.R. Tyson, S.P. Garbett, B. Weidow, M.P. Harris, W. Georgescu, Trait variability of cancer cells quantified by high-content automated microscopy of single cells. Meth. Enzymol. **467**, 23–57 (2009)
13. D.K. Singh, C.-J. Ku, C. Wichaidit, R.J. Steininger, L.F. Wu, S.J. Altschuler, Patterns of basal signaling heterogeneity can distinguish cellular populations with different drug sensitivities. Mol. Syst. Biol. **6**, 369 (2010)
14. M.D. Slack, E.D. Martinez, L.F. Wu, S.J. Altschuler, Characterizing heterogeneous cellular responses to perturbations. Proc. Natl. Acad. Sci. USA **105**(49), 11–19306 (2008)
15. B. Snijder, L. Pelkmans, Origins of regulated cell-to-cell variability. Nat. Rev. Mol. Cell Biol. **12**(2), 25–119 (2011)
16. M. Stamatakis, K. Zygourakis, A mathematical and computational approach for integrating the major sources of cell population heterogeneity. J. Theor. Biol. **266**(1), 41–61 (2010)

17. D. Stockholm, R. Benchaouir, J. Picot, P. Rameau, T.M.A. Neildez, G. Landini et al., The origin of phenotypic heterogeneity in a clonal cell population in vitro. PLoS ONE **2**(4), e394 (2007)
18. C. Tomelleri, E. Milotti, C. Dalla Pellegrina, O. Perbellini, A. Del Fabbro, M.T. Scupoli, A quantitative study of growth variability of tumour cell clones in vitro. *Cell Prolif. 2008th ed.* **41**(1), 177–191 (2008)
19. D.R. Tyson, S.P. Garbett, P.L. Frick, V. Quaranta, Fractional proliferation: a method to deconvolve cell population dynamics from single-cell data. *Nat. Methods 2012th ed.* **9**, 923–928 (2012)
20. J.J. Tyson, B. Novak, Regulation of the eukaryotic cell cycle: molecular antagonism, hysteresis, and irreversible transitions. J. Theor. Biol. **210**(2), 63–249 (2001)
21. H. Youssefpour, X. Li, A.D. Lander, J.S. Lowengrub, Multispecies model of cell lineages and feedback control in solid tumors. J. Theor. Biol. **304**, 39–59 (2012)

# A Stochastic Model of Glioblastoma Invasion: The Impact of Phenotypic Switching

**Philip Gerlee and Sven Nelander**

**Abstract** In this chapter we present a stochastic model of glioblastoma (brain cancer) growth and invasion, which incorporates the notion of phenotypic switching between migratory and proliferative cell states. The model is characterised by the rates at which cells switch to proliferation ($q_p$) and migration ($q_m$), and simulation results show that for a fixed $q_p$, the tumour growth rate is maximised for intermediate $q_m$. We also complement the simulations by deriving a continuum description of the system, in the form of two coupled reaction-diffusion PDEs, and subsequent phase space analysis shows that the wave speed of the solutions closely matches that of the stochastic model. The model thus reveals a possible way of treating glioblastomas by altering the balance between proliferative and migratory behaviour.

**Keywords** Brain tumour · Cell-based model · Travelling wave analysis

## 1 Introduction

Tumour growth is dependent on numerous intra-cellular and extra-cellular processes, such as an elevated rate of proliferation, evasion of apoptosis and angiogenesis [5]. Out of these, proliferation has traditionally been singled out as the most important, and has generally been the target of anti-cancer therapies. However, recently there has been a growing interest in the impact of cancer cell motility, since it underlies

---

P. Gerlee (✉)
Department of Integrated Mathematical Oncology, H. Lee Moffitt Cancer Center and Research Institute, Tampa, USA
e-mail: philip.gerlee@moffitt.org

S. Nelander
Department of Immunology, Genetics and Pathology and Science for Life Laboratory, Uppsala University, Uppsala, Sweden
e-mail: sven.nelander@gu.se

M. Delitala and G. Ajmone Marsan (eds.), *Managing Complexity, Reducing Perplexity*,
Springer Proceedings in Mathematics & Statistics 67, DOI: 10.1007/978-3-319-03759-2_3,
© Springer International Publishing Switzerland 2014

the invasive nature of tumour growth. This process is especially relevant in the case of glioblastoma, which generally exhibit diffuse morphologies stemming from the high motility of individual glioma cells. Recent experimental work suggests that migration and proliferation in glioma cells are mutually exclusive phenotypes [2], where the cells move in a saltatory fashion interspersed by periods of stationary behaviour during which cell division occurs. In this chapter we explore the theoretical implications of this observations using a stochastic individual-based model. In particular we are interested in how the rates of phenotypic switching (microscopic parameters) influence the growth rate of the tumour (a macroscopic property).

Before proceeding to the model description let us briefly mention that glioblastoma has a long history of mathematical modelling dating back to the seminal work of Murray et al. (see for example [8], Chap. 11 for an in-depth review), who employed a continuous reaction-diffusion approach. Recently stochastic and individual-based models have gained in popularity and several such models have been proposed [1, 3, 6, 7].

## 2 Stochastic Model

The cells are assumed to occupy a $d$-dimensional lattice with lattice spacing $\Delta x$ (we will consider $d = 1, 2$), containing $N^d$ lattice sites, where $N$ is the linear size of the lattice and each lattice site either is empty or holds a single glioma cell. This means that we disregard the effects of the surrounding brain tissue, such as the different properties of grey versus white matter [9], and the presence of capillaries which might influence the behaviour of the cancer cells. For the sake of simplicity we do not consider any interactions between the cancer cells (adhesion or repulsion), although this could easily be included.

The behaviour of each cell is modelled as a time continuous Markov process, where each transition or action occurs with a certain rate. Each cell is assumed to be in either of two states: proliferating or migrating, and switching between the states occurs at rates $q_p$ (into the P-state) and $q_m$ (into the M-state). A proliferating cell is stationary, passes through the cell cycle, and thus divides at a rate $\alpha$. The daughter cell is placed with uniform probability in one of the empty $2d$ neighbouring lattice sites (using a von Neumann neighbourhood). If the cell has no empty neighbours cell division fails. A migrating cell performs a size exclusion random walk, where each jump occurs with rate $v$. Size exclusion means that the cell can only move into lattice sites which were previously empty.

Lastly, cells are assumed to die, of natural causes, at a rate $\mu$ independent of the cell state, after which they are removed from the lattice and leave an empty lattice site behind. The stochastic process is depicted schematically in Fig. 1, and the model parameters are summarised in Table 1.

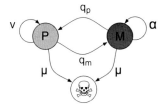

**Fig. 1** Schematic describing the continuous time Markov process each cell in the model follows. A living cell can be in either of two states, proliferating (P) or migrating (M), and transitions between the states with rates $q_p$ and $q_m$ respectively. A P-cell divides at rate $\alpha$ while an M-cell moves with rate $v$. Both cell types die with a constant rate $\mu$.

**Table 1** Summary of model parameters. All rates are given in units of cell cycle$^{-1}$

| Meaning | Name | Value |
|---|---|---|
| Rate of switching to P-state | $q_p$ | 0–30 |
| Rate of switching to M-state | $q_m$ | 0–30 |
| Proliferation rate | $\alpha$ | 1 |
| Motility rate | $v$ | 5 |
| Death rate | $\mu$ | $10^{-3}$ |
| Lattice spacing | $\Delta x$ | $20\,\mu\mathrm{m}$ |

## 3 Simulation Results

A typical simulation outcome is displayed in Fig. 2a, which shows the spatial distribution of tumour cells after $T = 50$ cell cycles have passed. The initial condition was a single cell in the proliferative state at the centre of the lattice, and the phenotypic switching rates were set to $(q_p, q_m) = (20, 10)$. This plot gives us a general idea of the dynamics of the model; the tumour grows with a radial symmetry, and exhibits a solid core, while the tumour margin is diffuse and somewhat rugged. However, in order to get a wider picture of the influence of the phenotypic switching rates on tumour mass, we measured the number of cancer cells at $T = 50$ in the parameter range $0 < q_{p,m} < 30$. The result of this parameter sweep is displayed in Fig. 2b and shows a strong influence of the two parameters. For $q_p = 0$ all cells are in the migratory state and hence the tumour does not grow at all, while the other extreme $q_m = 0$ gives rise to compact tumours driven purely by cell division. These results are intuitive, but what is more interesting is that tumour cells with intermediate switching rates are the ones that give rise to the largest tumours. Although migratory behaviour does not directly contribute to an increase in the number of cancer cells it has the secondary effect of freeing up space which accelerates growth compared to the tumours dominated purely by proliferation ($q_m = 0$). The results suggest that for each $q_p > 0$ there is a $q_m \neq 0$ which gives rise to a maximal tumour mass.

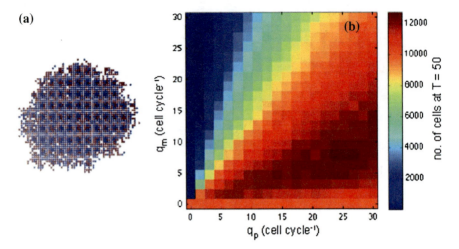

**Fig. 2** Growth dynamics of the model. **a** Shows the result of a single simulation of the model for $(q_p, q_m) = (10, 20)$ while **b** Shows the average tumour mass as a function of the phenotypic switching rates

## 4 Continuum Approximation

The counter-intuitive results of the previous section spurred us to investigate the dynamics of the model from an analytical perspective. We will here give a brief outline of an attempt employing a continuum approximation which gives an estimate of the tumour interface velocity as function of the model parameters. For a full account of the derivation we refer the reader to [4].

By considering the processes which affect the cells on the lattice (proliferation, movement, phenotypic switching and death), and by assuming independence of the lattice sites we can derive master equations for the occupation probabilities of P- and M-cells. By taking the appropriate continuum limit we arrive at the following system of coupled PDEs which describe the density of P- and M-cells respectively:

$$\frac{\partial p}{\partial t} = \frac{\alpha}{2}(1 - p - m)\frac{\partial^2 p}{\partial x^2} + \alpha p(1 - p - m) - (q_m + \mu)p + q_p m \quad (1)$$

$$\frac{\partial m}{\partial t} = \frac{v}{2}((1 - p)\frac{\partial^2 m}{\partial x^2} + m\frac{\partial^2 p}{\partial x^2}) - (q_p + \mu)m + q_m p. \quad (2)$$

Despite its seeming complexity this system bears resemblance to the Fisher equation [8], and similarly exhibits travelling wave solutions. It is the propagation speed $c$ of these solutions, that correspond to the rate of invasion, which we are hoping to determine. With the travelling wave ansatz ($z = x - ct$) the above system is turned into the following system of autonomous ODEs:

# A Stochastic Model of Glioblastoma Invasion: The Impact of Phenotypic Switching

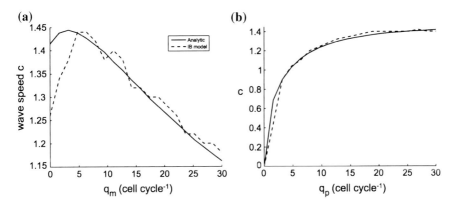

**Fig. 3** Comparison of the wave speed obtained from the phase space analysis (*solid line*) and the one observed in simulation of the stochastic model (*dashed line*). In **a** the switch rate to proliferation is fixed at $q_p = 15$, while in **b** we have fixed $q_m = 15$

$$\begin{aligned}
P' &= Q \\
M' &= N \\
Q' &= \tfrac{2}{\alpha(1-P-M)}((q_m + \mu)P - q_p M - cQ - \alpha P(1 - P - M)) \\
N' &= \tfrac{2}{v(1-P)}((q_p + \mu)M - \tfrac{vM}{\alpha(1-P-M)}((q_m + \mu)P \\
&\quad - q_p M - cQ - \alpha P(1 - P - M)) - cN - q_m P)
\end{aligned}$$

with boundary conditions

$$\begin{aligned}
P(-\infty) &= p^\star \quad M(-\infty) = m^\star \quad Q(-\infty) = 0 \quad N(-\infty) = 0 \\
P(\infty) &= 0 \quad\quad M(\infty) = 0 \quad\quad Q(\infty) = 0 \quad\quad N(\infty) = 0
\end{aligned} \quad (3)$$

where $(p^\star, m^\star)$ corresponds to the stable invaded state of Eq. (1)–(2) and (0, 0) is the unstable healthy state. As in the case of the Fisher equation we find the speed of propagation as the smallest $c$ for which the heteroclinic orbit connecting the unstable and stable state remains non-negative for all times [8]. In our case this boils down to a four-dimensional eigenvalue problem involving the Jacobian of the system (3), which unfortunately does not have a closed form solution. However by fixing the model parameters a numerical solution can easily be found.

Figure 3 shows a comparison between the wave speed obtained from the phase space analysis and the one obtained from simulating the stochastic model. It is clear that the analytical wave speed agrees well with the one observed in simulation, and also that least agreement occurs for small $q_m$ when the contribution of diffusive behaviour to growth is small.

Naturally the other parameters of the model also influence the rate of invasion. By fixing the phenotypic switching rates at $(q_p, q_m) = (10, 10)$ and by varying the other parameters ($\alpha$, $v$, and $\mu$) independently the curves in Fig. 4 were obtained. From these it can be seen that for small proliferation rates we have $c \sim \sqrt{\alpha}$, and for all motility

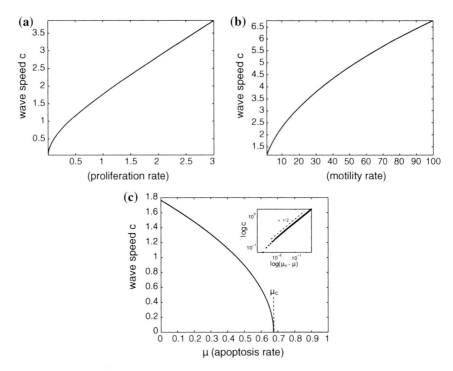

**Fig. 4** The wave speed of the propagating tumour margin as a function of **a** $\alpha$, **b** $v$ and **c** $\mu$. The dashed line in the inset of **c** has slope 1/2 and shows that $c \sim (\mu_c - \mu)^{1/2}$

rates in the range we see that $c \sim \sqrt{v}$. The death rate has a negative impact on the wave speed, and interestingly it seems as if the system goes through a second-order phase-transition, since above some critical $\mu_c$ the wave speed is equal to zero, and that it approaches this point with a diverging derivative $dc/d\mu$. Upon closer inspection we observed that $c \sim (\mu_c - \mu)^\beta$, with the critical exponent $\beta = 0.5049 \pm 0.0004$ being independent of the other parameters, while $\mu_c$ is parameter specific.

## 5 Discussion

From our simulations and analysis it is obvious that if glioma cells are engaged in phenotypic between migratory and proliferative behaviour, then the rates at which this occurs has a strong impact on tumour growth rate. In particular we have shown that for each $q_p > 0$ there exists a $q_m \neq 0$ which maximises the tumour interface velocity. A simple explanation of the influence the switching rates have on tumour growth velocity, is that they change the geometry and structure of the tumour interface, which in turn alters its growth velocity. A one-dimensional growth process in which the tumour expands in a narrow channel will suffice for the illustration.

If $q_m = 0$, then the tumour expands only through proliferation of the cells at the interface (since interior cells cannot divide), and the interface thus moves with velocity $\alpha$. If $q_m \neq 0$ then cells at the interface spend some time in the motile state and, with non-zero probability, move away from the tumour mass, freeing up space and thus allowing previously blocked cells to proliferate. This process increases the interface velocity, but it is also clear that for large $q_m$ the velocity is lowered, since if $q_m \gg q_p$ few cells are in the proliferative state and can thus take advantage of the space created by migrating cells. From this perspective it is clear that the tumour interface velocity will depend on $q_m$ in a non-monotone way, and in fact the phase space analysis shows that for each $q_p \neq 0$ the velocity $c = c(q_m)$ attains a maximum, which occurs at $q_m^{max} \approx 0.5 q_p$.

Despite its apparent theoretical nature the model could, if properly parametrised, give indications as to the efficacy of certain therapies. It could serve as a tool mapping perturbations at the cellular level caused by a drug to the impact those changes have on tumour growth rate. If a drug for example influences the rates of phenotypic switching, then it could potentially both increase and decrease tumour growth rate, depending on where in parameter space the unperturbed cells are located. It is believed that many drugs have precisely this dual impact on both proliferation and migration, and estimating the tissue-level of these perturbations effect will be difficult, if not impossible without the use of mathematical models such as this one.

# References

1. M. Aubert, M. Badoual, C. Christov, B. Grammaticos, A model for glioma cell migration on collagen and astrocytes. J. R. Soc. Interface **5**(18), 75 (2008)
2. A. Farin, S.O. Suzuki, M. Weiker, J.E. Goldman, J.N. Bruce, P. Canoll, Transplanted glioma cells migrate and proliferate on host brain vasculature: a dynamic analysis. Glia **53**(8), 799–808 (2006). doi:10.1002/glia.20334
3. S. Fedotov, A. Iomin, Probabilistic approach to a proliferation and migration dichotomy in tumor cell invasion. Phys. Rev. E **77**(3), 31911 (2008)
4. P. Gerlee, S. Nelander. The impact of phenotypic switching on glioblastoma growth and invasion. PLoS Comput. Biol. **8**(6), e1002556 (2012)
5. D. Hanahan, R.A. Weinberg, The hallmarks of cancer. Cell **100**(1), 57–70 (2000)
6. H. Hatzikirou, D. Basanta, M. Simon, K. Schaller, A. Deutsch, Go or grow: the key to the emergence of invasion in tumour progression? Math. Med. Biol. **29**(1), 49–65 (2012)
7. E. Khain, M. Katakowski, S. Hopkins, A. Szalad, X. Zheng, F. Jiang, M. Chopp, Collective behavior of brain tumor cells: the role of hypoxia. Phys. Rev. E **83**(3), 031920 (2011). doi:10.1103/PhysRevE.83.031920
8. J.D. Murray, *Mathematical Biology II: Spatial Models and Biomedical Applications* (Springer, Verlag, 1989)
9. K.R. Swanson, E.C. Alvord, J.D. Murray, A quantitative model for differential motility of gliomas in grey and white matter. Cell Prolif **33**(5), 317–330 (2000)

# A Hybrid Model for *E. coli* Chemotaxis: From Signaling Pathway to Pattern Formation

**Franziska Matthäus**

**Abstract** In this article a hybrid model for the chemotactic motion of *E. coli* is presented that captures a description of the internal signaling pathway as well as the interaction of the bacteria with the surrounding ligand. The hybrid nature of the model arises from the fact that discrete agents interact with and through an external chemoattractant that is described as a continuous variable. Motion of the bacteria is not restricted to the numerical grid on which the chemoattractant concentration is defined. Local production and uptake of ligand allow a study of the effects of internal signaling processes on pattern formation processes or on the fitness of populations in competition for a common nutrient source. This model provides a tool to connect individual-based models to continuous (PDE) descriptions for bacterial chemotaxis.

**Keywords** Chemotaxis · Signaling pathway · Pattern formation · Multi-scale mathematical model

## 1 Introduction

In this chapter we present a hybrid model for *E. coli* motion. The model is given as a cellular automaton, providing a description of internal signaling processes of *E. coli*, coupled with a continuous description (PDE) for the dynamics of external chemical substances. The model allows to simulate movements of individual cells as well as large-scale population behavior. It provides a tool to study bacterial pattern formation processes or competition of different species under the influence of internal signaling processes.

---

F. Matthäus (✉)
Center for Modeling and Simulation in the Biosciences (BIOMS), University of Heidelberg,
Im Neuenheimer Feld 294, 69120 Heidelberg, Germany
e-mail: franziska.matthaeus@iwr.uni-heidelberg.de

M. Delitala and G. Ajmone Marsan (eds.), *Managing Complexity, Reducing Perplexity*,
Springer Proceedings in Mathematics & Statistics 67, DOI: 10.1007/978-3-319-03759-2_4,
© Springer International Publishing Switzerland 2014

The choice of *E. coli* bacteria as a model organism is straight forward. Its chemotaxis signaling pathway is very simple and well understood, and several mathematical models have been developed to describe the pathway [1, 2, 10]. The chemotaxis signaling pathway of *E. coli* connects a membrane receptor to the flagella. The flagella can either rotate clockwise or counter-clockwise. Counter-clockwise rotation thereby causes a (more or less) straight swim, while clockwise rotation leads to a so-called tumble, where a new direction is chosen without translational movement. The receptor switches between two states, active and inactive. This switching is random, but influenced by the external ligand concentration. If the receptor is active, it phosphorylates an enzyme CheA, which in turn phosphorylates CheY. Phosphorylated CheY then binds to the flagellar motor and induces tumbling. A feedback loop involving a further enzyme, CheB, introduces memory. CheB is activated by CheAp and, together with its antagonist, CheR, is involved in receptor methylation. Methylation also influences the receptor's probability to be in the active state, and counteracts the effects of the ligand. Through this process the bacteria are able to adapt to constant ligand concentrations, and to compare present ligand concentration to past values. This "chemical memory" is needed for chemotaxis, if the ligand concentration during a run of the bacterium increases, the the tumble probability decreases, and vice versa.

There exist several agent-based models describing *E. coli* motion subject to internal signaling processes. AgentCell [8] relies on a stochastic simulation of the enzymatic interactions (StochSim [11]) and is therefore computationally expensive for larger bacterial populations. The model of Bray [3, 4] describes the signaling pathway in terms of about 90 differential equations, and accounts also for processes like receptor assembly. Also this model is not suitable for simulating large populations. Vladimirov et al. [17] and Curk et al. [6] developed coarse-grained models capturing the essential behavior of the signaling pathway without details on the enzymatic reactions. These models, on the other hand, are very suitable for large-scale population studies, but do not allow to study the specific influence of signaling pathway processes on the macroscopic behavior. None of these models accounts for an interaction of the bacteria with the ligand.

Here, we will extend a model previously developed to study the motion of *E. coli* bacteria in various chemical landscapes [12]. The model was used to investigate the influence of noise in the signaling pathway on the random search behavior [13] and the chemotactic precision. The signaling pathway is thereby described as a small system of differential equations, comprising an equation for the $m$-times methylated receptor, and the enzymes CheAp, CheBp, and CheYp. Two algebraic equations describe the probability of the receptor to be in the active state, and the tumbling probability depending on the concentration of CheYp. In the following sections we will describe the model and the extension that describes interaction with the ligand (production, uptake). First simulations with the extended model show a chemotactic pattern formation process for a small number of bacteria.

# 2 Methods

## 2.1 ODE System Describing the Signaling Pathway

The internal signaling pathway of the bacteria is described as a system of ordinary differential equations, adapted from [10]. We denote the $m$-times methylated receptor by $T_m$, and the concentration of phosphorylated form of CheA, CheB and CheY by $A_p$, $B_p$ and $Y_p$, respectively.

$$\frac{dT_m}{dt} = k_R R \frac{T_{m-1}}{K_R + T^T} + k_B B_p \frac{T^A_{m+1}}{K_B + T_A} - k_R R \frac{T_m}{K_R + T^T} \tag{1a}$$

$$- k_B B_p \frac{T^A_m}{K_B + T_A} \tag{1b}$$

$$\frac{dA_p}{dt} = k_A (A^T - A_p) T_A - k_Y A_p (Y^T - Y_p) - k'_B A_p (B^T - B_p) \tag{1c}$$

$$\frac{dY_p}{dt} = k_Y A_p (Y^T - Y_p) - k_Z Y_p Z - \gamma_Z Y_p \tag{1d}$$

$$\frac{dB_p}{dt} = k'_B A_p (B^T - B_p) - \gamma_B B_p \tag{1e}$$

The model is extended by two algebraic equations: $p_m(L)$ describes the probability of the $m$-times methylated receptor to be active under a given ligand concentration $L$:

$$p_m(L) = V_m \left( 1 - \frac{L^{H_m}}{L^{H_m} + K_m^{H_m}} \right). \tag{2}$$

The last equation connects the concentration of CheYp to the tumbling probability $\tau$:

$$\tau = \frac{Y_p^{H_c}}{Y_p^{H_c} + K_c^{H_c}}. \tag{3}$$

For the values of the parameters see [12]. The tumbling probability depends on the internal concentration levels of all enzymes involved in the chemotaxis signaling pathway, and on the external ligand concentration. Bacterial trajectories are generated from the output variable ($\tau$) in the following way. While swimming, the bacteria preserve the direction. During a tumble, a new direction is chosen randomly, following a $\Gamma$-distribution with shape parameter 4, scale parameter 18.32 and location $-4.6$ [8].

## 2.2 Ligand Dynamics

To model the interaction of the bacteria with the ligand, we take a square domain $\Omega = [x_0, x_{max}] \times [y_0, y_{max}]$ with zero-flux or periodic boundary conditions. For numerical treatment, the domain is discretized, and the ligand concentration $L_{i,j}$ is given at the grid points only. Ligand diffuses with diffusion coefficient $D_L$, and is degraded with rate $r_L$. Ligand may also be produced locally in space, specified by a function $L_{prod}(x, y)$. Production of ligand by the bacteria is given by the function $L_{bact}(p)$, where $p$ denotes a set of parameters, for instance related to the internal enzyme concentration. Uptake of the ligand by the bacteria is given by the rate $r_{bact}$:

$$ r_{bact} = k_u \cdot \frac{L}{L + K}, \tag{4} $$

with the constants $k_u$ and $K$. The dependence Eq. (4) ensures that uptake of ligand is proportional to the ligand concentration, but no larger than a maximum value $k_u$.

## 2.3 Hybrid Model for the Interaction of the Bacteria with the Ligand

Since the run length and turning angles of the bacteria are random, their movement is not restricted to the numerical grid on which the ligand concentration is defined. In our model, the bacteria interact with the four nearest grid points. If the distance between two grid points is chosen to be one unit, the four surrounding grid points of a bacterium located at $(x, y)$ are given by $\{(\text{int}(x),\text{int}(y)), (\text{int}(x),\text{int}(y)+1), (\text{int}(x)+1,\text{int}(y)), (\text{int}(x)+1,\text{int}(y)+1)\}$.

## 2.4 Interaction Weights

The strength of the interaction between the bacterium and a grid point depends on the distance. The smaller the distance the larger the interaction. With $dx$ and $dy$ as shown in Fig. 1, we measure the distance to a grid point by the maximum norm, which is natural in grid-environments. The unnormed interaction weights between the bacterium and the four grid points are given as $\tilde{w}_{i,j} = 1 - ||d_{i,j}||_\infty$, which turns out to be

$$ \begin{aligned} \tilde{w}_{i,j} &= \min(1 - dx, 1 - dy) \quad \tilde{w}_{i+1,j} = \min(dx, 1 - dy) \\ \tilde{w}_{i,j+1} &= \min(1 - dx, dy) \quad \tilde{w}_{i+1,j+1} = \min(dx, dy). \end{aligned} \tag{5} $$

The sum of the unnormed interaction weights still depends on $dx$ and $dy$. The final interaction weights will be normed and given by $w_{i,j} = \tilde{w}_{i,j} / \sum \tilde{w}$. With this definition the four weights add up to 1 for any pair $dx$ and $dy$.

**Fig. 1** Distance of the bacterium to the surrounding grid points

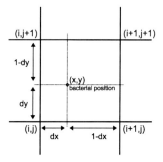

## 3 Results

We tested the described setting by modeling the chemotactic motion of a small number of *in silico* bacteria to self-produced ligand. We chose a small quadratic domain of length $X = 125\,\mu\text{m}$. The diffusion coefficient of the ligand was set to $D_L = 0.015\,\mu\text{m}^2/\text{s}$, and its degradation rate to $r_L = 0.001\,\mu\text{M/s}$. Bacteria produced ligand locally with rate $L_{\text{bact}}(p) = \text{const} = 1\,\mu\text{M/s}$. These parameters deviate from the physiological parameters, but are chosen such that a small number of bacteria are able to produce detectable gradients. Physiological parameters would be $D_L \approx 1 \times 10^{-4}\,\text{mm}^2/\text{s}$ and a smaller production rate, which, however, would require a much larger number of cells and a significantly larger domain to generate concentration profiles that allow for pattern formation processes.

### 3.1 Computational Cost

The computational cost of the simulations arises on one hand from the numeric solution of Eqs. (1a–3) for every bacterium. On the other hand, also the operations on the ligand (especially diffusion) adds to the cost. For a scaling of the computation time with the domain size $X$, number of bacteria $N$ and the simulated time interval $T$ see Fig. 2. Simulations of 1,000 bacteria for 20 min (natural generation time) and a milimeter-scale domain size would still be feasible with the present setting.

### 3.2 Formation of Transient Patterns

With the given parameters the bacteria produce detectable gradients. The simulated bacteria often turn and retrace their own path. They also produce local ligand accumulations by moving in a very confined area for a certain time. In most of the simulations, the bacteria accumulate in a small area and thus produce a local maximum in

| $T$ [min] | $N$ | comp. cost (time [min]) |
|---|---|---|
| 2 | 1 | 1.62 |
| 2 | 10 | 2.34 |
| 2 | 100 | 9.59 |
| 10 | 1 | 8.08 |
| 10 | 10 | 11.65 |
| 10 | 100 | 49.24 |

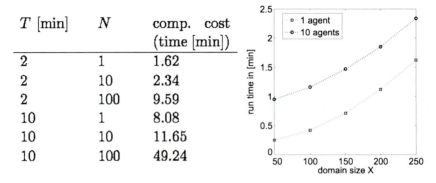

**Fig. 2** Computational cost for simulations of the motion of $N$ bacteria during a time interval of length $T$ (in [$min$]) in a domain of size $X = 250\,\mu$m (*left*) and for an increasing domain size for 1 and 10 bacteria respectively and a simulated time of T = 2 min (*right*)

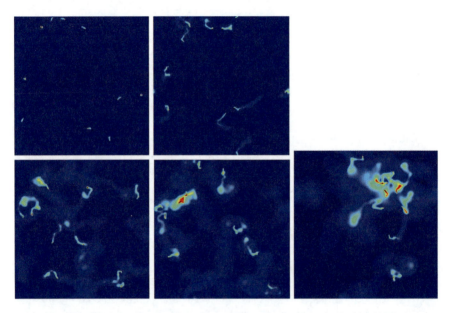

**Fig. 3** *Snapshots of bacterial motion* Snapshots of 10 moving bacteria, producing chemoattractant. The sequence on the *left* shows the emergence of a local accumulation in the *top left* corner. The figure on the *right* another aggregation involving 9 out of 10 bacteria

the ligand concentration to which they respond chemotactically, as shown in Fig. 3. These accumulations are, however, transient. The transient nature probably arises from a combination of adaptation to the absolute ligand concentration and stochastic effects caused by the small number of bacteria. Whether the accumulation pattern can be stabilized (for larger domain and population sizes, and different parameters of ligand diffusion, production and degradation) will be the topic of further studies.

## 4 Discussion and Outlook

The presented model framework allows simulation of *E. coli* motion and chemotaxis for large populations under consideration of detailed aspects of the chemotaxis signaling pathway. The framework can therefore be seen as a tool to connect models of the signaling pathway and agent-based approaches of (like the models of Emonet et al. [8] and Zonia and Bray [18]) to models considering pattern formation processes on the population density scale (i.e. models described by Hillen and Painter [9], Polezhaev et al. [15] or Tyson et al. [16]). While the influence of signaling pathway processes on motion behavior has been well studied for single individuals, there are only very few studies on pattern formation on the population that include signal processes [7, 14]. Most approaches include signal processing only in a very phenomenological way. In fact, *E. coli* is the only organism where the enzymatic reactions comprising the chemotaxis signaling pathway are understood to this detail.

The presented framework allows the study of pattern formation processes of mutants, or of individuals affected by noise in the signaling pathway. Also competition of different populations for a common nutrient source can be simulated when including ligand uptake. Already in first test simulations, transient aggregations can be produced, which are have been observed experimentally and in simulations for *E. coli* bacteria swimming in liquid medium (see for instance [5, 16]). Extensions and modifications of the model might also enable the reproduction of more common patterns in semi-solid medium.

**Acknowledgments** This work was supported by the Center for Modeling and Simulation in the Biosciences (BIOMS), University of Heidelberg.

## References

1. S. Asakura, H. Honda, Two-state model for bacterial chemoreceptor proteins. J. Mol. Biol. **176**, 349–367 (1984)
2. N. Barkai, S. Leibler, Robustness in simple biochemical networks. Nature **387**, 913–917 (1997)
3. D. Bray, R.B. Burret, M.I. Simon, Computer simulation of the phosphorylation cascade controlling bacterial chemotaxis. Mol. Biol. Cell **4**, 469–482 (1993)
4. D. Bray, M.D. Levin, K. Lipkow, The chemotactic behavior of computer-based surrogate bacteria. Curr. Biol. **17**(4), R132–R134 (2007)
5. E.O. Budrene, H.C. Berg, Complex patterns formed by motile cells of *E. coli*. Nature **349**, 630–633 (1991)
6. T. Curk, F. Matthäus, Y. Brill-Karniely, J. Dobnikar. in *Advances in Systems Biology, Advances in Experimental Medicine and Biology*, ed. by I.I. Goryanin, A.B. Goryachev. Chapter coarse

graining *E. coli* chemotaxis: from multi-flagella propulsion to logarithmic sensing, vol. 736 (Springer, New York, 2011)

7. J.C. Dallon, H.G. Othmer, A discrete cell model with adaptive signaling for aggregation of dictyostelium discoideum. Phil. Trans. R. Soc. Lond. B **352**, 391–417 (1997)
8. T. Emonet, C.M. Macal, M.J. North, C.E. Wickersham, P. Cluzel, Agentcell: a digital single-cell assay for bacterial chemotaxis. Bioinformatics **21**(11), 2714–2721 (2005)
9. T. Hillen, K.J. Painter, A user's guide to pde models for chemotaxis. J. Math. Biol. **58**(1), 183–217 (2009)
10. M. Kollmann, L. Løvdok, K. Bartholomé, J. Timmer, V. Sourjik, Design principles of a bacterial signalling network. Nature **438**, 504–507 (2005)
11. N. Le Novère, T.S. Shimizu, Stochsim: modelling of stochastic biomolecular processes. Bioinformatics **6**(6), 575–576 (2001)
12. F. Matthäus, M. Jagodić, J. Dobnikar, *E. coli* superdiffusion and chemotaxis—search strategy, precision and motility. J. Biophys **97**(4), 946–957 (2009)
13. F. Matthäus, M.S. Mommer, T. Curk, J. Dobnikar, On the origin and characteristics of noise-induced lévy walks of *E. coli*. PLOS One **6**(4), e18623 (2011)
14. E. Palsson, H.G. Othmer. A model for individual and collective cell movement in dictyostelium discoideum. PNAS **97**(19), 10448–10453 (2000)
15. A. Polezhaev, R.A. Pashkov, A.I. Lobanov, I.B. Petrov, Spatial patterns formed by chemotactic bacteria *E. coli*. Int. J. Dev. Biol. **50**, 309–314 (2006)
16. R. Tyson, S.R. Lubkin, J.D. Murray, A minimal mechanism for bacterial pattern formations. Proc. R. Soc. Lond. B **266**, 299–304 (1999)
17. N. Vladimirov, L. Løvdok, D. Lebiedz, V. Sourjik, Dependence of bacterial chemotaxis on gradient shape and adaptation rate. PLOS Comp. Biol **4**(12), e1000242 (2008)
18. L. Zonia, D. Bray, Swimming patterns and dynamics of simulated escherichia coli bacteria. J. R. Soc. Interface **6**, 1035–1046 (2009)

# Multiscale Analysis and Modelling for Cancer Growth and Development

**Dumitru Trucu and Mark A. J. Chaplain**

**Abstract** In this chapter we present a novel framework that enables a rigorous analysis of processes occurring on three (or more) independent scales (e.g. intracellular, cellular, tissue). We give details of the establishment of this new multiscale concept and discuss a number of important fundamental properties that follow. This framework also offers a new platform for the analysis of a new type of multiscale model for cancer invasion that we propose. This new model focuses on the macroscopic dynamics of the distributions of cancer cells and of the surrounding extracellular matrix and its connection with the microscale dynamics of the matrix degrading enzymes, produced at the level of the individual cancer cells.

**Keywords** Multiscale modelling · Cancer invasion · Computational simulations

## 1 Introduction

Cancer growth is a complex process that develops over several spatial and temporal scales, ranging from genes to molecular, cellular, and tissue levels. The spatial multiscale character plays a crucial part in the overall tumour development and is present from the very early stages when avascular solid tumours are formed. Characterised by a diffusion-limited growth, these avascular solid tumours have a final size of about 2 mm in diameter ($10^9$ cells) consisting of an inner necrotic core, a middle quiescent region, and an outer proliferating rim. During the invasive phase of their growth, tumour cells produce matrix degrading enzymes (MDEs), such as the matrix metalloproteinases (MMPs) [9], which are secreted into the extracellular

---

D. Trucu (✉) · M. A. J. Chaplain
Division of Mathematics, University of Dundee, Dundee DD1 4HN, UK
e-mail: trucu@maths.dundee.ac.uk

M. A. J. Chaplain
e-mail: chaplain@maths.dundee.ac.uk

M. Delitala and G. Ajmone Marsan (eds.), *Managing Complexity, Reducing Perplexity*, 45
Springer Proceedings in Mathematics & Statistics 67, DOI: 10.1007/978-3-319-03759-2_5,
© Springer International Publishing Switzerland 2014

matrix (ECM) via a dynamic process of growth of receptors bound to the cancer cell membrane. As a consequence, this ability of cancer cells to break out of tissue compartments and spread locally, gives solid tumours a defining deadly characteristic and is a crucial step in the process of metastasis [25]. However, it is important to observe the genuinely spatial multiscale perspective of the overall cancer invasion. In the micro-scale stage of the invasion process, the ECM degradation is caused by the evolving spatial distributions of secreted MDEs and occurs at a molecular/cell level. Once the matrix is degraded, the cancer invades the tissue at a macroscopic level.

Understanding the many processes involved in cancer cell invasion of tissue is therefore of great importance for gaining a deeper insight into cancer growth and development, and the design of future anti-cancer strategies. Over the last two decades, there has been a great effort in characterising the cancer invasion process via mathematical modelling, see for example [4, 5, 7, 11–13, 17, 20, 24]. Along these concerted modelling and analytical approaches, the multiscale character of cancer invasion has already been recognised as being an essential part in the overall invasion process and debated in various regards, see [6, 8, 14, 18–20].

Developments have also been taking place within the multiscale area, both from an analytical and a numerical stand point, see [1, 2, 10, 15, 21–23]. These pave the way for a deeper understanding and more rigorous formulation of the processes occurring on three (or more) distinct scales: namely the intracellular scale (inside the cell), the intercellular scale (between cells), and the tissue scale. Generally speaking, one may refer to these scales as the microscale, the mesoscale and the macroscale. Therefore, we will naturally have two scaling factors $\lambda > 0$ and $\sigma > 0$ that realise the transition between the macro- and meso-scale and meso- and micro-scale, respectively. As explored in great analytical detail in [22], the multiscale character of cancer invasion as well as various other multiscale questions arising in material science or soft-matter physics has generated interest in the establishment of a multiscale framework that is able to deal with more than three scales, when the scaling factors $\lambda$ and $\sigma$ are not functions of the same reference parameter, say $\varepsilon > 0$. From a mathematical stand point, it is usual that this kind of activity on three-scales can be described asymptotically by a family of partial differential or integral operators $\mathscr{L}_{\lambda,\sigma}$, whose coefficients are dependent on the microscale $\sigma$ and mesoscale $\lambda$, which, for a given domain $\Omega$, under the presence of appropriate initial and boundary conditions, captures the underlying complex process in terms of a corresponding family of solutions $u_{\lambda,\sigma}$ that is obtained for the induced systems of equations:

$$\mathscr{L}_{\lambda,\sigma} = f. \tag{1}$$

Thus, the solutions $u_{\lambda,\sigma}$ of Eq. (1) inherently depend on the micro-, meso- and macro-scales. Depending on the particularities of each process and the heterogeneous medium under investigation, one may consider whether to adopt a macroscale approximation of the process via a homogenization approaches (if this is possible and appropriate) or to perform another type of asymptotic analysis. In the next section

Multiscale Analysis and Modelling for Cancer Growth

we will describe the new notion of three-scale convergence that offers a platform for introducing a new multiscale topology in which such three-scale processes could be assessed.

## 2 The Concept of Three-Scale Convergence

While a certain notion of multiscale convergence has previously been introduced by Allaire and Briane in Ref. [3], obtained in essence by iterating the two-scale convergence defined by [16], we will focus our attention on defining and exploring a new concept of three-scale convergence where the scaling factors $\lambda$ and $\sigma$ are independent in the sense that they are not functions of a common reference parameter $\varepsilon$. In order to introduce this multiscale concept, let us first proceed with a few notations. Let us consider $\Omega$ a bounded region in $\mathbb{R}^N$ and let $Y := [0, 1]^N$ be the unit cube. Let us denote by $\mathscr{C}_{\#}^{\infty}(Y)$ the set of infinitely differentiable functions on $\mathbb{R}^N$ obtained as a Y—periodical extension of $\mathscr{C}^{\infty}(Y)$. Further, $H_{\#}^1(Y)$ will denote the completion of $\mathscr{C}_{\#}^{\infty}(Y)$ for the norm of $H^1(Y)$. Also, let us consider the space $\mathscr{D}(\Omega; \mathscr{C}_{\#}^{\infty}(Y; \mathscr{C}_{\#}^{\infty}(Y)))$ that consists of all test functions $\psi(x, y, z)$ having the properties that, for any fixed $x$, the function $\psi(x, \cdot, \cdot)$ belongs to $\mathscr{C}_{\#}^{\infty}(Y; \mathscr{C}_{\#}^{\infty}(Y))$. Finally, if we fix an arbitrary $y$, the function $\psi(x, y, \cdot)$ belongs to $\mathscr{C}_{\#}^{\infty}(Y)$. For any two sets of indices $\Sigma, \Lambda \subset \mathbb{R}$ that accumulate to zero, under the previous notations, the properties of the three-scale convergence concept, introduced and explained in full details in [22], are reviewed here in brief as follows:

**Definition 1** A sequence of functions $\{u_{\lambda,\sigma}\}_{\lambda \in \Lambda, \sigma \in \Sigma} \subset L^2(\Omega)$ is said to be three-scale convergent to a function $u_0 \in L^2(\Omega \times Y \times Y)$ if, for any $\psi \in \mathscr{D}(\Omega; \mathscr{C}_{\#}^{\infty}(Y; \mathscr{C}_{\#}^{\infty}(Y)))$, denoting

$$\lim_{\Lambda, \Sigma} \int_{\Omega} u_{\lambda,\sigma}(x) \psi(x, \frac{x}{\lambda}, \frac{x}{\lambda\sigma}) dx := \lim_{\Lambda} \left[ \lim_{\Sigma} \int_{\Omega} u_{\lambda,\sigma}(x) \psi(x, \frac{x}{\lambda}, \frac{x}{\lambda\sigma}) dx \right], \quad (2)$$

the following relation holds true:

$$\lim_{\Lambda, \Sigma} \int_{\Omega} u_{\lambda,\sigma}(x) \psi(x, \frac{x}{\lambda}, \frac{x}{\lambda\sigma}) dx = \iiint_{\Omega \times Y \times Y} u_0(x, y, z) \psi(x, y, z) dx dy dz. \quad (3)$$

The well-posedness of the new concept of three-scale convergence is justified as follows.

**Theorem 1** *From any arbitrary* $\| \cdot \|_{L^2(\Omega)}$*-bounded sequence* $\{u_{\lambda,\sigma}\}_{\lambda \in \Lambda, \sigma \in \Sigma} \subset L^2$ $(\Omega)$ *we can extract a subsequence that is three-scale convergent to a limit* $u_0 \in$ $L^2(\Omega \times Y \times Y)$.

Further, the boundedness properties of a three scale convergence sequence is explored by the following result.

**Theorem 2** *Let $\{u_{\lambda,\sigma}\}_{\lambda \in \Lambda, \sigma \in \Sigma} \subset L^2(\Omega)$ be a three-scale convergent sequence to a function $u_0 \in L^2(\Omega \times Y \times Y)$. Then there exists a constant $M > 0$ as well as two particular indices $\lambda_0 \in \Lambda$ and $\sigma_0 \in \Sigma$ such that, for $(\lambda, \sigma) \in \Lambda \times \Sigma$ with $\lambda \leq \lambda_0$ and $\sigma \leq \sigma_0$, we have*

$$\| u_{\lambda,\sigma} \|_{L^2(\Omega)} \leq M. \tag{4}$$

The following theorem gives a compactness characterisation for a product of sequences that are three-scale convergent. This is similar to the notion of "strong convergence" encountered in two-scale convergence, see Ref. [2].

**Theorem 3** *Let $\{u_{\lambda,\sigma}\}_{\lambda \in \Lambda, \sigma \in \Sigma} \subset L^2(\Omega)$ be a sequence that is three-scale convergent to a function $u_0 \in L^2(\Omega \times Y \times Y)$, which satisfies the following property:*

$$\lim_{\Lambda, \Sigma} \| u_{\lambda,\sigma} \|_{L^2(\Omega)} = \| u_0 \|_{L^2(\Omega \times Y \times Y)}. \tag{5}$$

*Then, for any sequence $\{v_{\lambda,\sigma}\}_{\lambda \in \Lambda, \sigma \in \Sigma}$ that three-scale converges to $v_0 \in L^2(\Omega \times Y \times Y)$, we have*

$$u_{\lambda,\sigma} v_{\lambda,\sigma} \rightharpoonup \iint_{Y \times Y} u_0(x, y, z) v_0(x, y, z) dy dz \quad in \ \mathscr{D}'(\Omega). \tag{6}$$

Finally, the convergence of the gradients is obtained via the following theorem.

**Theorem 4** *Let $\{u_{\lambda,\sigma}\}_{\lambda \in \Lambda, \sigma \in \Sigma} \subset H^1(\Omega)$ be a bounded sequence with respect to $\| \cdot \|_{H^1(\Omega)}$. Then, there exist three functions*

$$\begin{aligned} u_0 &\in H^1(\Omega), \\ u_1 &\in L^2(\Omega, H^1_\#(Y)), \\ u_2 &\in L^2(\Omega \times Y, H^1_\#(Y)), \end{aligned} \tag{7}$$

*and a subsequence $\{u_{\lambda,\sigma}\}_{\lambda \in \bar{\Lambda}, \sigma \in \bar{\Sigma}} \subset \{u_{\lambda,\sigma}\}_{\lambda \in \Lambda, \sigma \in \Sigma}$ such that we have:*

1. $\{u_{\lambda,\sigma}\}_{\lambda \in \bar{\Lambda}, \sigma \in \bar{\Sigma}}$ *is three-scale convergent to $u_0$;*
2. $\{\nabla u_{\lambda,\sigma}\}_{\lambda \in \bar{\Lambda}, \sigma \in \bar{\Sigma}}$ *is three-scale convergent to $\nabla u_0 + \nabla_y u_1 + \nabla_z u_2$.* $\tag{8}$

## 3 A Three-Scale Process Arising in a Multiscale Moving Boundary Model for Cancer Invasion

In the cancer invasion context, let us assume a simplified picture in which we are concerned only with the dynamics of the ECM and cancer cells that are located within a maximal reference spatial cube $Y \subset \mathbb{R}^N (N = 2, 3)$, which is centred at the origin 0. Given $\varepsilon$, where $0 < \varepsilon < 1$, we will consider an $\varepsilon$—resolution of $Y$, i.e. a uniform decomposition of $Y$ using spatially translated $\varepsilon Y$ cubes. Let $t_0$ be an arbitrarily chosen time. In the cancer affected region $\Omega(t_0)$, the macro-dynamics of $c_{\Omega(t_0)}(x, t)$ and $v_{\Omega(t_0)}(x, t)$ occurring over the time interval $[t_0, t_0 + \Delta t]$ are governed by the following coupled macro-process. Firstly, the equation governing the cancer cell population consists of a diffusion term as well as a term modelling the directed haptotactic movement to the ECM, along with a term describing cancer cell proliferation, i.e.

$$\frac{\partial c_{\Omega(t_0)}}{\partial t} = D \Delta c_{\Omega(t_0)} - \eta \nabla \cdot (c_{\Omega(t_0)} \nabla v_{\Omega(t_0)}) + g(c_{\Omega(t_0)}, v_{\Omega(t_0)}). \tag{9}$$

The equation governing the ECM concentration consists of a degradation term in the presence of the cancer cells along with a general remodelling term, i.e.

$$\frac{d v_{\Omega(t_0)}}{dt} = -\alpha(t) c_{\Omega(t_0)} v_{\Omega(t_0)} + \zeta(c_{\Omega(t_0)}, v_{\Omega(t_0)}), \tag{10}$$

where $\alpha(t)$ is a homogeneous time-dependent degradation factor.

The macro-process described by Eqs. (9–10) have the following initial conditions:

$$\begin{aligned} c_{\Omega(t_0)}(x, t_0) &=: c^0_{\Omega(t_0)}(x), \quad x \in \Omega(t_0), \\ v_{\Omega(t_0)}(x, t_0) &=: v^0_{\Omega(t_0)}(x), \quad x \in \Omega(t_0), \end{aligned} \tag{11}$$

as well as certain moving boundary conditions that are imposed by the microscopic dynamics arising within a $\varepsilon$—bundle $\mathscr{P}_\varepsilon$ of $\varepsilon$—size cubes $\varepsilon Y$ that cover $\partial \Omega(t_0)$, namely $\partial \Omega(t_0) \subset \bigcup_{\varepsilon Y \in \mathscr{P}_\varepsilon} \varepsilon Y$.

In brief, these $\varepsilon$—cubes are chosen so that, on one hand, one face is captured inside $\Omega(t_0)$, which we denote by $\Gamma^{int}_{\varepsilon Y}$. On the other hand, the faces that are perpendicular on $\Gamma^{int}_{\varepsilon Y}$ are all intersecting $\partial \Omega(t_0)$ while the face that is parallel to $\Gamma^{int}_{\varepsilon Y}$ is remaining completely outside of the only connected component of $\varepsilon Y \cap \Omega(t_0)$ that is containing $\Gamma^{int}_{\varepsilon Y}$. On each of these micro-domains $\varepsilon Y$, an MDE micro-dynamics takes place. Since the MDEs are secreted locally by the cancer cells from within $\Omega(t_0)$, for any $\tau \in [0, \Delta t]$, the local mean-value of the cancer cells spatial distribution $c_{\Omega(t_0)}(\cdot, t_0 + \tau)$ can be considered to describe the source for the degrading enzymes within $\varepsilon Y \cap \Omega(t_0)$.

Therefore, on each micro-domain $\varepsilon Y \in \mathscr{P}_\varepsilon$, we obtain a space-wise compact support source $f_{\varepsilon Y} : \varepsilon Y \times [0, \Delta t] \to \mathbb{R}_+$ such that, for any $\tau \in [0, \Delta t]$, $f_{\varepsilon Y}(\cdot, \tau)$ has the following properties:

1. $f_{\varepsilon Y}(y, \tau) = \frac{1}{\lambda(\mathbf{B}(y, 2\varepsilon) \cap \Omega(t_0))} \int_{\mathbf{B}(y, 2\varepsilon) \cap \Omega(t_0)} c_{\Omega(t_0)}(x, t_0 + \tau) dx, \quad y \in \varepsilon Y \cap \Omega(t_0),$

2. $f_{\varepsilon Y}(y, \tau) = 0, \quad y \in \varepsilon Y \setminus \left(\Omega(t_0) + \{z \in Y \mid \|z\|_2 < \gamma\}\right),$ (12)

where $\lambda(\cdot)$ is the standard Lebesgue measure on $\mathbb{R}^N$, $\mathbf{B}(y, 2\varepsilon) := \{x \in Y \mid \|y - x\|_\infty \leq 2\varepsilon\}$, and $\gamma$ is a constant parameter chosen such that $\gamma \ll \frac{\varepsilon}{3}$. Hence, denoting by $m(y, t)$ the MDE distribution on $\varepsilon Y$, during the time period $[0, \Delta t]$, on any $\varepsilon Y \in \mathscr{P}_\varepsilon$, the rate of change of the matrix degrading enzyme molecular distribution per unit time is modelled as the effect of a diffusion process under the presence of the source term $f_{\varepsilon Y}(y, \tau)$, i.e.

$$\frac{\partial m}{\partial \tau} = \Delta m + f_{\varepsilon Y}(y, \tau), \quad y \in \varepsilon Y, \ \tau \in [0, \Delta t],$$ (13)

with zero initial conditions and zero Neumann boundary conditions.

Denoting by $x_{\varepsilon Y}^*$ the first point of the intersection between the median of $\Gamma_{\varepsilon Y}^{int}$ and $\partial \Omega(t_0)$, of great interest is the possible displacement of $x_{\varepsilon Y}^*$ to a new spatial location $\widetilde{x_{\varepsilon Y}^*}$ as a result of the micro-process that is taking place on $\varepsilon Y$. This displacement occurs when a certain transitional probability $q(x_{\varepsilon Y}^*) := \frac{1}{\int_{\varepsilon Y} m(y, \Delta t) dy} \int_{\varepsilon Y \setminus \Omega(t_0)}$ $m(y, \Delta t) dy$ exceeds a certain spatially associated threshold $\omega_{\varepsilon Y} \in (0, 1)$ that is induced by the local characteristics of the tissue confined within $\varepsilon Y$.

If the threshold $\omega_{\varepsilon Y}$ is exceeded, a third spatial scale is used to described the pattern in the MDEs distribution $m(y, \Delta t)$ that ultimately determines the displacement direction as well as its magnitude. This third scale is obtained via the regularity property of the Lebesgue measure, and it is given by the maximal resolution size of the uniform dyadic decomposition $\{\mathscr{D}_j\}_{j \in \mathscr{J}}$ that can be accepted for $\varepsilon Y$ such that the non-overlapping region $\varepsilon Y \setminus \Omega(t_0)$ is approximated with accuracy $\delta \ll \varepsilon$ by the union of the sub-family of dyadic cubes

$$\{\mathscr{D}_i\}_{i \in \mathscr{J}_\delta} := \{\mathscr{D} \in \{\mathscr{D}_j\}_{j \in \mathscr{J}} \mid \mathscr{D} \subset \varepsilon Y \setminus \Omega(t_0)\}.$$ (14)

If we denote this scale by $\sigma$, it is immediate to remark that for two different micro-domains $\varepsilon Y$s the associated $\sigma$—scales will be different, and as a consequence in this three-scale process, the $\varepsilon$—scale and $\sigma$—scale remain independent.

This $\sigma$—scale is used to define the direction of potential movement and the magnitude of potential displacement, but for conciseness purposes, in this presentation we do not enter in the details of how these displacement characteristics are derived, and, for full explanations, we refer the reader to [23].

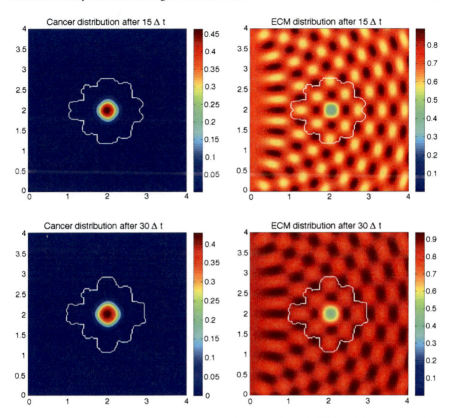

**Fig. 1** Plots showing the computed distributions of cancer cells and ECM, and the contours of the invasive boundary of the invading tumour after both 15 and the 30 macro-micro invasion stages

Therefore, a new boundary $\partial \Omega(t_0 + \Delta t)$ will be obtained as a smooth interpolation of the set consisting of the new spatial positions obtained for those $x^*_{\varepsilon Y}$ that were moved and of the existing spatial positions of the rest of the $x^*_{\varepsilon Y}$ that were not moved. Further, the macroscopic dynamics on the new domain $\Omega(t_0 + \Delta t)$ will continue to be defined by the same governing Eqs. (9–11), but having the new initial conditions determined by the solution at the final time of the previous invasion macro-step that is smoothly extended on the difference region $\Omega(t_0 + \Delta t) \setminus \Omega(t_0)$ via a convolution process with a fast-decaying compact support kernel. For full details of this new multiscale moving boundary model as well as the new multiscale numerical scheme it gave rise to and computational invasion results that were obtained, we refer the reader to [23].

In order to explore this multiscale model computationally, we developed a novel multiscale numerical technique. Briefly, this numerical approach combines a finite difference approximation of the macro-dynamics with a finite element scheme used for the boundary micro-dynamics. The computational simulation results in Fig. 1 show the spatio-temporal evolution of cancer cells and ECM alongside the invasive tumour boundary and reveals a pronounced lobular and fingered-type progression typical of cancer invasion patterns.

## 4 Concluding Remarks

While the process of cancer growth and development presents us with a vast range of multiscale sub-processes, with various independent scales, the new concept of three-scale convergence is paving the way for the establishment of an analytical framework that will be appropriate for rigorous investigation. A concrete example of such a three-scale process arises within cancer invasion, where a certain built-in stochasticity that appears at the lower scales determines the cancer cell invasion pathways in the surrounding ECM. The proposed multiscale modelling approach is able to reveal a pronounced heterogeneous progression of cancer invasion (lobular and fingered protrusions into the ECM). The computational simulation of this model has led to the development of a novel type of multiscale "front-checking" numerical scheme, whose robustness and consistency properties will be investigated within the three-scale framework.

**Acknowledgments** MAJC and DT gratefully acknowledge the support of the European Research Council through the ERC AdG Grant 227619 *From Mutations to Metastases: Multiscale Mathematical Modelling of Cancer Growth and Spread.*

## References

1. A. Abdulle, C. Schwab, Heterogeneous multiscale FEM for diffusion problems on rough surfaces. Mult. Mod. Sim. **3**, 195–220 (2004)
2. G. Allaire, Homogenization and two-scale convergence SIAM. J. Math. Anal. **23**, 1482–1518 (1991)
3. G. Allaire, M. Briane, Multiscale convergence and reiterated homogenization Proc. Roy. Soc. Edin. **126A**, 297–342 (1996)
4. V. Andasari, A. Gerisch, G. Lolas, A.P. South, M.A.J. Chaplain, Mathematical modeling of cancer cell invasion of tissue: biological insight from mathematical analysis and computational simulation. J. Math. Biol. **63**, 141–171 (2011)
5. A.R.A. Anderson, K.A. Rejniak, P. Gerlee, V. Quaranta, Microenvironment driven invasion: a multiscale multimodel investigation. J. Math. Biol. **58**, 579–624 (2009)
6. N. Bellomo, E. De Angelis, L. Preziosi, Multiscale modelling and mathematical problems related to tumor evolution and medical therapy. J. Theor. Med. **5**, 111–136 (2004)
7. M.A.J. Chaplain, G. Lolas, Mathematical modelling of cancer cell invasion of tissue: The role of Ethe urokinase plasminogen activation system. Math. Mod. Meth. Appl. Sci. **15**, 1685–1734 (2005)
8. TS. Deisboeck, Z. Wang, P. Macklin, V Cristini, Multiscale Cancer Modeling. Annu. Rev. Biomed. Eng. **13** (2011)
9. M. Egeblad, Z. Werb, New functions for the matrix metalloproteinases in cancer progression. Nat. Rev. Cancer **21**, 161–174 (2002)

10. B. Engquist, Y.-H. Tsai, Heterogeneous multiscale methods for stiff ordinary differential equations. Math. Comp. **74**(252), 1707–1742 (2005)
11. H.B. Frieboes, X. Zheng, C.H. Sun, B. Tromberg, R. Gatenby, V. Christini, An integrated computational/experimental model of tumor invasion. Cancer Res. **66**, 1597–1604 (2006)
12. R.A. Gatenby, E.T. Gawlinski, A reaction-diffusion model of cancer invasion. Cancer Res. **56**, 5745–5753 (1996)
13. A. Gerisch, M.A.J. Chaplain, Mathematical modelling of cancer cell invasion of tissue: ELocal and non-local models and the effect of adhesion. J. Theor. Biol. **250**, 684–704 (2008)
14. M. Lachowicz, Micro and meso scales of description corresponding to a model of tissue invasion by solid tumours. Math. Mod. Meth. Appl. Sci. **15**, 1667–1683 (2005)
15. P. Lin, Convergence analysis of a quasi-continuum approximation for a two-dimensional material without defects. SIAM J. Num. Anal. **45**, 313–332 (2007)
16. G. Nguetseng, A general convergence result for a functional related to the theory of homogenization. SIAM J. Math. Anal. **20**, 608–623 (1989)
17. A.J. Perumpanani, J.A. Sherratt, J. Norbury, H.M. Byrne, Biological inferences from a mathematical model for malignant invasion. Invasion Metastasis **16**, 209–221 (1996)
18. L. Preziosi, A. Tosin, Multiphase and multiscale trends in cancer modelling. Math. Model. Nat. Phenom. **4**, 1–11 (2009)
19. L. Preziosi, A. Tosin, Multiphase modelling of tumour growth and extracellular matrix interaction: mathematical tools and applications. J. Math. Biol. **58**, 625–656 (2009)
20. I. Ramis-Conde, D. Drasdo, A.R.A. Anderson, M.A.J. Chaplain, Modeling the Influence of the E-Cadherin-$\beta$-Catenin Pathway in Cancer Cell Invasion: a Multiscale Approach. Biophys. J. **95**, 155–165 (2008)
21. W.Q. Ren, E. Weinan, Heterogeneous multiscale method for the modeling of complex fluids and micro-fluidics. J. Comp. Phys. **204**, 1–26 (2005)
22. D. Trucu, M.A.J. Chaplain, A. Marciniak-Czochra, Three-scale convergence for processes in heterogeneous media. Appl. Anal. Int. J. (2011). DOI: 10.1080/00036811.2011.569498
23. D. Trucu, P. Lin, M.A.J. Chaplain, Y. Wang, A multiscale moving boundary model arising in cancer invasion. SIAM Multiscale Model. Simul. J. **11**, 309–335 (2013)
24. S. Turner, J.A. Sherratt, Intercellular adhesion and cancer invasion: a discrete simulation using the extended Potts model. J. Theor. Biol. **216**, 85–100 (2002)
25. D. Hanahan, R.A. Weinberg, The hallmarks of cancer. Cell **100**, 57–70 (2000)

# A Non-linear Flux-Limited Model for the Transport of Morphogens

**J. Calvo, J. Soler and M. Verbeni**

**Abstract** Morphogenic proteins play a key role in developmental biology. We introduce flux-limited diffusion as a new tool to obtain mathematical descriptions of biological systems whose fate is controlled by this class of proteins.

**Keywords** Sonic Hedgehog (Shh) pathway · Morphogen propagation · Reaction-diffusion equations · Finite propagation speed · Flux-limited diffusion equations · Flux-saturated diffusion equations

## 1 Introduction

Morphogenic proteins are main protagonists in crucial aspects of developmental biology. Their importance comes from the fact that they mediate intercellular communication acting as signaling molecules. They are also related to tumorigenesis [29]. These proteins are usually issuing from localized sources in the extracellular medium, originating a concentration gradient. Target cells will respond to the instructive signals according to both their concentration and duration [15, 29]. The outcome is a change in gene transcription, which plays a pivotal role in generating cellular diversity and patterning. Cells need time to process the protein signals and to give a genetic response. Modelling these phenomena constitutes then a complex problem

J. Calvo (✉)
Departament de Tecnologies de la Informació i les Comunicacions, Universitat Pompeu Fabra, Barcelona, Spain
e-mail: juan.calvo@upf.edu

J. Soler · M. Verbeni
Departamento de Matemática Aplicada, Universidad de Granada, Granada, SpainJ. Soler
e-mail: jsoler@ugr.es

M. Verbeni
e-mail: mverbeni@ugr.es

M. Delitala and G. Ajmone Marsan (eds.), *Managing Complexity, Reducing Perplexity*,
Springer Proceedings in Mathematics & Statistics 67, DOI: 10.1007/978-3-319-03759-2_6,
© Springer International Publishing Switzerland 2014

in which different spatial and temporal scales are involved. Besides, the biological mechanisms of transport, reception and gradient formation of morphogenic signals are recently argued to be realized through cell extensions (nanotubes, filopodia or cytonemes) [4, 19, 28]. This opens a new perspective on the subject that revises the basis of the previous models of morphogenesis based on Brownian motion and then on linear diffusion principles.

In this work we focus on the mechanisms that regulate the whole chemical cascade in which morphogenic proteins are involved, with special emphasis on the associated space transport mechanisms. Here we will concentrate on the mathematical description of the Sonic Hedgehog (Shh hereafter) morphogenic function, whose task is to promote the expression of the Gli genetic code. The Shh/Gli code is involved in the development of the embryo, a biological system that has been thoroughly studied [15, 16, 20, 29, 30, 33].

The model case which has been most dealt with in the literature is that of a chick or mice embryo, in which measurements can be afforded [5, 15, 16, 20, 29]. More precisely, this morphogenic protein induces the dorsoventral patterning of the spinal cord in the neural tube, which is the precursor of the central nervous system. It must be pointed out that there is a privileged way of propagation in the neural tube, see [15, 16, 20, 29, 30, 33] (the so called dorsoventral axis, DV hereafter) and in such a way a convenient simplifying assumption is to regard the system as being one-dimensional along the DV axis. Another point which is worth mentioning is that there is an almost similar biological system, the so-called wing imaginal disc in Drosophila. Here the Hh morphogenic protein plays an analogous role to that of Shh in vertebrates and the Gli target gene for Shh has its counterpart in the Ci—cubitus interruptus—gene for Hh in drosophila [5, 8]. Being easier to perform measurements in this setting, this provides a powerful and handy workbench in order to try to describe in a more complete way the development of the neural tube.

## 2 Mathematical Models

### 2.1 Linear Diffusion Models

Morphogen propagation has been studied from the mathematical point of view since long ago, starting with the work of Alan Turing [32], which has been successively improved by a series of authors in different contexts, see the recent review [23].

The most accepted models up to date assume that the spreading of the morphogen is described with linear diffusion mechanisms, based on microscopic Brownian motion. Then, the standard models use reaction-diffusion equations. On one hand, linear diffusion equations are used to describe morphogen propagation and the formation of concentration gradients (Shh in our case). On the other hand, the law of mass action is used to describe the rates of change of the protein concentrations

A Non-linear Flux-Limited Model for the Transport of Morphogens

involved in the transduction of the Shh signal (GliA among others for the case we are interested in) and gene activations.

As the culmination of several decades of work we find the mathematical model by Saha and Schaffer [30]. Its main purpose is to understand how morphogen gradients are formed and interpreted from a dynamical point of view. This model studies DV patterning in the chick embryo spinal cord, beginning when Shh is first secreted by the floor plate (see for example [16]). It focuses not on the whole neural tube, but only on the ventral-most binary cell fate (V3 interneurons). The model consists of a reaction-diffusion equation for the spreading of the Shh morphogen

$$\frac{\partial[\text{Shh}]}{\partial t} = D_{\text{Shh}}\Delta[\text{Shh}] + k_{\text{off}}[\text{PtcShh}_{mem}] - k_{\text{on}}[\text{Shh}][\text{Ptc}_{mem}] \qquad (1)$$

(square brackets denote concentrations) plus a set of ordinary differential equations for the concentrations of the most relevant proteins involved in the transduction process: $\text{PtcShh}_{mem}$, $\text{PtcShh}_{cyt}$, $\text{Ptc}_{mem}$, $\text{Ptc}_{cyt}$, $\text{Gli1}^{Act}$, $\text{Gli3}^{Act}$ and $\text{Gli3}^{Rep}$. All the equations are posed on a finite spatial interval and (1) is complemented with Neumann boundary conditions at the left end and zero Dirichlet boundary conditions at the right end.

The main drawback that we find in this model is the unphysical spreading of the morphogen to all the neural tube soon after secretion, entailed by the presence of the Laplacian operator [33]. One of the mechanisms considered to deal with this situation is to take into account an (static) artificial activation threshold (for the Shh concentration), below which no chemical reactions take place, see Fig. 2c and [30]. This amounts to cut off *a posteriori* the numerical profiles obtained as solutions to (1), thus introducing artificial fronts (Fig. 2c). This is a very delicate issue, as several recent experimental findings point out. Namely, the concentration of Shh received by the cells and the time of exposure are factors of similar relevance [15, 16]. To sum up, without a threshold mechanism the chemical signal arrives too fast to distant areas, thus triggering the chemical cascade too soon. But with a threshold mechanism the chemical signal will never be able to activate distant cells (contrary to the long-range signaling effect that has been observed [30]), having as a result that large sections of the neural tube will never be exposed to the action of the morphogen. Apart form this, it has been also shown that Shh does not travel through the medium as it stands, but as a part of bigger aggregates or vesicles [8, 34, 35]. As the size of these aggregates is comparable to that of the medium through which they are moving, being also this medium quite inhomogeneous, the usual scale assumptions for a description in terms of Brownian motion are not fulfilled at all. Then we commit ourselves to give an alternative transport mechanism to the linear diffusion that is able to reproduce the recent experimental results, see Fig. 2b and [33].

We identify as the source of most of the problems the recourse to linear diffusion mechanisms, which is not realistic in this context. The basic issues would be then to remove the infinite speed of propagation, to allow for front propagation instead and also to account properly for the temporal and spatial scales involved in the process. Our proposal is to substitute the linear diffusion mechanisms and to use flux limitation instead. Then we have to deal with a non-linear flux-limited reaction-diffusion system [33], as we explain below. The first aim consist in obtaining a graded temporal distribution of the signal and, as a consequence, to recover the time necessary to activate or inhibit the different genes involved in the signal transduction. This is not allowed when the velocity of propagation is infinite, as the natural inhibitor-activator process requires some time to develop [25].

## 2.2 A Non-linear Flux-Limited Model

The problem of infinite speed of propagation for linear diffusion equations dates back to Fourier's theory of heat conduction [18], which he based on a linear relation between the heat flux and the gradient $\nabla u(t, x)$ of the temperature function. The subsequent macroscopic equation, $\partial_t u = k \Delta_x u$, predicts an infinite speed of propagation for the heat. Flux-limitation mechanisms propose to modify Fourier's law to obtain a saturating heat flow when temperature gradients become unbounded. A variety of macroscopic equations [2, 24, 26] are then produced, among which the following is a remarkable example:

$$\frac{\partial u}{\partial t} = \nu \text{div} \left( \frac{|u|^m \nabla_x u}{\sqrt{u^2 + \frac{\nu^2}{c^2} |\nabla_x u|^2}} \right) + F(u) . \tag{2}$$

In the case $m = 1$, this equation with $F(u) = 0$ was first introduced by Rosenau in [27] and later derived by means of optimal mass transportation in [7]. It can also be recovered performing macroscopic limits of kinetic models [6]. Here the constant $c$ is the maximum speed of propagation allowed in the medium (analogous to the sound speed in hyperbolic settings; in fact the behavior of this equation is more hyperbolic than parabolic), a fact which is analytically justified in [3]. Furthermore $\nu$ stands for a kinematic viscosity and reduces to a diffusion coefficient in the limit $c \to \infty$, in which the usual heat equation is recovered [11]. The mathematical properties of this equation and related models have been analyzed in a series of papers (see [12] and references therein). For the case $m > 1$ (a porous-media flux-limited equation) we refer to [2, 13, 14].

We wonder next if we can tackle the qualitative behavior that we have in mind using this family of non-linear mechanisms. To test such an issue, these tools were incorporated into a widely known model, the one-dimensional FKPP reaction-diffusion model describing traveling waves [17, 22], which consists of Eq. (2) with $F(u) = k_0 u (1 - u)$, where $k_0$ is a constant related to the intrinsic growth rate of the biological particles.

It is found that, while classical traveling waves still do exist at high speeds, these degenerate to singular traveling fronts as the wave speed lowers to the value of the constant $c$ [10]. These singular traveling waves consist in a discontinuous entropy

# A Non-linear Flux-Limited Model for the Transport of Morphogens

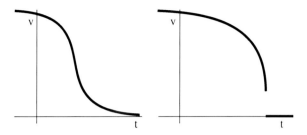

**Fig. 1** Different traveling waves: classical shape (*left*) and singular front (*right*)

solution with infinite tangent on the discontinuity front, see Fig. 1. Therefore, we learn that with these flux-limitation mechanisms we have a finite constant speed for the propagation of the biological information (whatever it may be) and the activation of related responses. Coming back to the description of morphogen propagation in the neural tube, the previous background encourages us to change the linear reaction-diffusion equation (1) describing transport of morphogens in the neural tube, introducing a flux limitation mechanism instead of the Laplace operator [33]. The following flux-limited spreading (FLS) equation results:

$$\frac{\partial [\text{Shh}]}{\partial t} = \nu \, \partial_x \left( \frac{[\text{Shh}] \partial_x [\text{Shh}]}{\sqrt{[\text{Shh}]^2 + \frac{\nu^2}{c^2}(\partial_x [\text{Shh}])^2}} \right) + k_{off}[\text{Ptc1Shh}_{mem}]$$
$$- k_{on}[\text{Shh}][\text{Ptc1}_{mem}].$$

Following this line of reasoning, the chemical reactions taking place inside the cells will be described by a set of ordinary differential equations different from those of [30], not only because the chemical signal does not arrive instantaneously to the surface receptors and this alters the internal dynamics in a significant way, but also because the synthesis and transport to cell membrane of Ptc1$_{cyt}$ molecules can take some time. This feature seems to have been overlooked in the previous models and it entails a delay for the system of differential equations (which is represented by the parameter $\tau$ below). The set of differential equations describing biochemical reactions inside the cells reads now as follows:

$$\frac{\partial [\text{Ptc1Shh}_{mem}]}{\partial t} = -(k_{\text{off}} + k_{\text{Cin}})[\text{Ptc1Shh}_{mem}] + k_{on}[\text{Shh}][\text{Ptc1}_{mem}]$$
$$+ k_{Cout}[\text{Ptc1Shh}_{cyt}],$$
$$\frac{\partial [\text{Ptc1Shh}_{cyt}]}{\partial t} = k_{\text{Cin}}[\text{Ptc1Shh}_{mem}] - k_{\text{Cout}}[\text{Ptc1Shh}_{cyt}] - k_{\text{Cdeg}}[\text{Ptc1Shh}_{cyt}],$$
$$\frac{\partial [\text{Ptc1}_{mem}]}{\partial t} = k_{\text{off}}[\text{Ptc1Shh}_{mem}] - k_{on}[\text{Shh}][\text{Ptc1}_{mem}] + k_{\text{Pint}}[\text{Ptc1}_{cyt}],$$
$$\frac{\partial [\text{Ptc1}_{cyt}]}{\partial t} = k_{\text{P}} P_{\text{tr}} \left\{ [\text{Gli1}^{Act}](t-\tau), [\text{Gli3}^{Act}](t), [\text{Gli3}^{Rep}](t) \right\} \Phi_{\text{Ptc}}$$

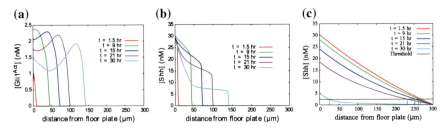

**Fig. 2** Evolution of Shh and Gli1$^{Act}$ versus distance from the floor plate at various times using our nonlinear flux-limited diffusion model in (**a**), (**b**). **c** represents the evolution of Shh when linear diffusion [30] is used, where a thershold and artificial fronts have been added. Note that in this case the fronts are moving backwards, while fronts moving forward appear in a natural way in our Gli-FLS system (**b**)

$$\frac{\partial [\text{Gli1}^{Act}]}{\partial t} = k_G P_{tr} \left\{ [\text{Gli1}^{Act}](t-\tau), [\text{Gli3}^{Act}](t), [\text{Gli3}^{Rep}](t) \right\} \Phi_{\text{Ptc}}$$
$$- k_{\text{Pint}}[\text{Ptc1}_{cyt}],$$
$$- k_{\text{deg}}[\text{Gli1}^{Act}],$$

$$\frac{\partial [\text{Gli3}^{Rep}]}{\partial t} = [\text{Gli3}^{Act}] \frac{k_{g3r}}{1 + R_{\text{Ptc}}} - k_{\text{deg}}[\text{Gli3}^{Rep}],$$

$$\frac{\partial}{\partial t}[\text{Gli3}^{Act}] = \frac{\gamma_{g3}}{1 + R_{\text{Ptc}}} - [\text{Gli3}^{Act}] \frac{k_{g3r}}{1 + R_{\text{Ptc}}} - k_{\text{deg}}[\text{Gli3}^{Act}],$$

being

$$\Phi_{\text{Ptc}} = \frac{[\text{Ptc1}_0]}{[\text{Ptc1}_0] + [\text{Ptc1}_{mem}]}, \quad R_{\text{Ptc}} = \frac{[\text{Ptc1Shh}_{mem}]}{[\text{Ptc1}_{mem}]},$$

where [Ptc1$_0$] is the initial value of [Ptc1$_{mem}$]. See [33] for the precise values of the parameters. From now on, we will refer to the coupling of the FLS equation with the ODEs system as the Gli-FLS model.

The mixed Dirichlet-Neumann problem (the well-possedness as well as the the asymptotic behavior of the solutions with zero weak Dirichlet boundary conditions at the right end [9]) for the FLS equation without reaction terms has been analyzed in [1, 9]. Interestingly enough, it is demonstrated that the incoming chemical signal travels exactly at constant speed $c$, which is precisely the behavior that we wanted to describe with a mathematical model, and which cannot be attained using a model like that in [30]. The value of $c$ can be measured experimentally in different systems [33]. The analysis of the complete Gli-FLS model is by now work in progress. We can nevertheless ascertain the behavior of our model by means of numerical simulations (see Fig. 2).

# A Non-linear Flux-Limited Model for the Transport of Morphogens

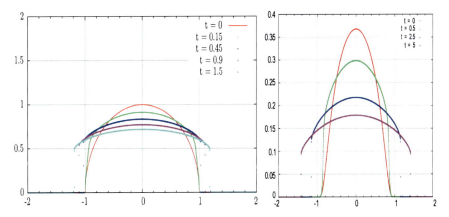

**Fig. 3** Evolution of two different initial data by Eq. (2) with $F = 0$. The first plot corresponds to $m = 4$, with $\sqrt{1-x^2}\chi_{|x|<1}$ as initial condition. For the second plot $m = 2$ and the initial condition is given by $\exp(1/(x^2 - 1))\chi_{|x|<1}$. In both cases $\nu = 1$ and $c = 1$. Area is preserved during evolution; the supports of the solutions start to grow when a vertical contact angle is reached

## 3 Comments and Discussion

We detail here the outstanding features of our proposed model. Using our equations we find that the unphysical diffusion of the Shh morphogen is eliminated. This entails the preservation of dynamical structures: the chemical signal is propagated as a traveling front, and now there are different biological responses at different times, instead of being activated instantaneously as it was the case with the linear diffusion model [30]. Our numerical simulations show a quite satisfactory agreement with experimental results [15, 33]. Cells can now be described as playing an active role in morphogen propagation and gradient formation. Besides, some suitable modifications and improvements which our model calls for are: the inclusion of Wnt, the possibility of describing also the relation with p53 [31], and the description of the BMP morphogenic family [21] (which is competing with the Shh in the development of the neural tube). We also keep in mind possible applications to cancer therapy. The first step would correspond to include the effect of the cyclopamine in our model. Another work in progress we can mention is the applicability of the porous-media flux-limited equation [2, 13] to this problem (see Fig. 3). As a final remark, our different approach could represent an important departure from the dominant interpretation of morphogenetic action modeling and understanding. While attractive for its apparent simplicity, linear diffusion cannot be the cornerstone to explain morphogenetic action. It may mimic biological patterns—which is not even the case for the situation that interests us here—, but it cannot account for how these are realized. Flux-limited diffusion may represent spatial constrains to morphogen movement through interaction with binding patterns or spreading through restricted channels.

**Acknowledgments** The Authors have been partially supported by Ministerio de Ciencia e Innovación (Spain), Project MTM2011-23384 and Junta de Andalucía Project P08-FQM-4267. J.C. is partially supported by a Juan de la Cierva fellowship.

# References

1. F. Andreu, J. Calvo, J.M. Mazón, J. Soler, On a nonlinear flux-limited equation arising in the transport of morphogenes. J. Diff. Eq. **250**(10), 5763–5813 (2012)
2. F. Andreu, V. Caselles, J.M. Mazón, J. Soler, M. Verbeni, Radially symmetric solutions of a tempered diffusion equation. A porous media-flux-limited case. SIAM J. Math. Anal. **44**, 1012–1049 (2012)
3. F. Andreu, V. Caselles, J.M. Mazón, S. Moll, Finite propagation speed for limited flux diffusion equations. Arch. Ration. Mech. Anal. **182**, 269–297 (2006)
4. P. Aza-Blanc, F.A. Ramírez-Webber, T.B. Kornberg, Cytonemes: cellular processes that project to the principal signaling center in drosophila imaginal discs. Cell **97**, 599–607 (1999)
5. P.A. Beachy, S.G. Hymowitz, R.A. Lazarus, D.J. Leahy, C. Siebold, Interactions between hedgehog proteins and their binding partners come into view. Genes Dev. **24**, 2001–2012 (2010)
6. N. Bellomo, A. Bellouquid, J. Nieto, J. Soler, Multiscale derivation of biological tissue models and flux-limited chemotaxis from binary mixtures of multicellular growing systems. Math. Models Methods Appl. Sci. **20**(7), 1179–1207 (2010)
7. Y. Brenier, in *Optimal Transportation and Applications*, ed. by L.A. Caffarelli, S. Salsa. Extended Monge-Kantorovich Theory. Lectures given at the C.I.M.E. Summer School help in Martina Franca. Lecture Notes in Math, vol. 1813 (Springer, Heidelberg, 2003), pp. 91–122
8. A. Callejo, A. Bilioni, E. Mollica, N. Gorfinkiel, G. Andrés, C. Ibáñez, C. Torroja, L. Doglio, J. Sierra, I. Guerrero, Dispatched mediates hedgehog basolateral release to form the long-range morphogenetic gradient in the drosophila wing disk epithelium. PNAS **108**(31), 12591–12598 (2011)
9. J. Calvo, J.M. Mazón, J. Soler, M. Verbeni, Qualitative properties of the solutions of a nonlinear flux-limited equation arising in the transport of morphogens. Math. Mod. Meth. Appl. Sci. **21**(supp1), 893–937 (2011)
10. J. Campos, P. Guerrero, O. Sánchez, J. Soler, On the analysis of travelling waves to a nonlinear flux limited reaction-diffusion equation. Annales de l'Institut Henri Poincaré (C) Analyse Non Linéaire **30**(1), 141–155 (2013)
11. V. Caselles, Convergence of the 'relativistic' heat equation to the heat equation as $c \to \infty$. Publ. Mat. **51**, 121–142 (2007)
12. V. Caselles, An existence and uniqueness result for flux limited diffusion equations. Discrete Continuous Dyn. Syst. **31**(4), 1151–1195 (2011)
13. V. Caselles, On the entropy conditions for some flux limited diffusion equations. J. Differ. Equ. **250**, 3311–3348 (2011)
14. A. Chertok, A. Kurganov, P. Rosenau, Formation of discontinuities in flux-saturated degenerate parabolic equations. Nonlinearity **16**, 1875–1898 (2003)
15. E. Dessaud, L.L. Yang, K. Hill, B. Cox, F. Ulloa, A. Ribeiro, A. Mynett, B.G. Novitch, J. Briscoe, Interpretation of the sonic hedgehog morphogen gradient by a temporal adaptation mechanism. Nature **450**, 717–720 (2007)
16. E. Dessaud, A.P. McMahon, J. Briscoe, Pattern formation in the vertebrate neural tube: a sonic hedgehog morphogen-regulated transcriptional network. Development **135**, 2489–2503 (2008)
17. R.A. Fisher, The advance of advantageous genes. Ann. Eugen. **7**, 335–369 (1937)
18. J. Fourier, Théorie Analithique de la Chaleur (1822)
19. H.H. Gerdes, R.N. Carvalho, Intercellular transfer mediated by tunneling nanotubes. Curr. Opin. Cell. Biol. **20**(4), 470–475 (2008)

20. P.W. Ingham, A.P. McMahon, Hedgehog signaling in animal development: paradigms and principles. Genes Dev. **15**, 3059–3087 (2001)
21. K. Kicheva, P. Pantazis, T. Bollenbach, Y. Kalaidzidis, T. Bittig, F. Jülicher, M. Gonzalez-Gaitan, Kinetics of morphogen gradient formation. Science **315**, 521 (2007)
22. A.N. Kolmogorov, I.G. Petrovsky, N.S. Piskunov, Etude de l'equation de la difusion avec croissance de la quantite de matiere et son application a un probléme biologique, Bulletin Universite d'Etata Moscou (Bjul. Moskowskogo Gos. Univ.). Serie internationale **A1**, 1–26 (1937)
23. S. Kondo, T. Miura, Reaction-diffusion model as a framework for understanding biological pattern formation. Science **329**, 1616–1620 (2010)
24. C.D. Levermore, G.C. Pomraning, A flux-limited diffusion theory. Astrophys. J. **248**, 321–334 (1981)
25. H. Meinhardt, Space-dependent cell determination under the control of a morphogen gradient. J. Theor. Biol. **74**, 307–321 (1978)
26. D. Mihalas, B. Mihalas, Foundations of radiation hydrodynamics (Oxford University Press, Oxford, 1984)
27. P. Rosenau, Tempered diffusion: a transport process with propagating front and inertial delay. Phys. Rev. A **46**, 7371–7374 (1992)
28. S. Roy, F. Hsiung, T.B. Kornberg, Specificity of drosophila cytonemes for distinct signaling pathways. Science **15**, 354–358 (2011)
29. A. Ruiz i Altaba, Combinatorial Gli gene function in floor plate and neuronal inductions by Sonic hedgehog. Development **125**, 2203–2212 (1998)
30. K. Saha, D.V. Schaffer, Signal dynamics in Sonic hedgehog tissue patterning. Development **133**, 889–900 (2006)
31. B. Stecca, A GLI1-p53 inhibitory loop controls neural stem cell and tumour cell numbers. EMBO J. **28**, 663–676 (2009)
32. A.M. Turing, The chemical basis of morphogenesis. Philos. Trans. R. Soc. Lond. Ser. B Biol. Sci. **237**, 37–72 (1952)
33. M. Verbeni, O. Sánchez, E. Mollica, I. Siegl-Cachedenier, A. Carleton, I. Guerrero, A. Ruiz i Altaba, J. Soler, Morphogen modeling through flux-limited spreading. Phys. Life Rev. **10**, 457–475 (2013)
34. J.P. Vincent, Hedgehog nanopackages ready for dispatch. Cell **133**, 1139 (2008)
35. N. Vyas, D. Goswami, A. Manonmani, P. Sharma, H. Ranganath, K. Vijay Raghavan, L. Shashidhara, R. Sowdhamini, S. Mayor, Nanoscale organization of hedgehog is essential for long-range signaling. Cell **133**, 1214 (2008)

# Glycosylation: A Phenomenon Shared by All Domains of Life

**Anne Dell and Federico Sastre**

**Abstract** This chapter provides insights into why proteins are glycosylated and how their glycosylation can be characterized by mass spectrometry. The covalent attachment of carbohydrates to proteins during their biosynthesis is a phenomenon shared by all domains of life. Indeed the majority of proteins in living systems are glycosylated. Their carbohydrates play critical roles in a myriad of biological processes especially those involving recognition. They do this via engagement with carbohydrate binding proteins called lectins. For example mammalian sperm-egg engagement in the first step of fertilization involves carbohydrate-lectin recognition, and the human egg is coated with a carbohydrate sequence called sialyl Lewisx which also plays important recognition roles in the immune system.

**Keywords** Biological recognition · Carbohydrates · Glycoproteins · Lectins · Mass spectrometry · Sialyl Lewisx

## 1 An Overview

The genome sequencing projects of the past two decades have yielded many surprises, the most startling of which is unquestionably the revelation that the total number of genes in humans is not very different from many model organisms such as worms, fruit flies and simple plants. This discovery has cast a spotlight on the correlate that biological complexity is not linearly related to the number of genes among species. Why might this be the case? A variety of explanations can be offered, arising from different fields of biological research. For example, molecular biologists might suggest transcriptional regulation or epigenetic modifications as key factors. Others

---

A. Dell (✉) · F. Sastre
Division of Molecular Biosciences, Faculty of Natural Sciences, Centre for Integrative Systems Biology, Imperial College London, London SW7 2AZ, UK
e-mail: a.dell@imperial.ac.uk

M. Delitala and G. Ajmone Marsan (eds.), *Managing Complexity, Reducing Perplexity*, Springer Proceedings in Mathematics & Statistics 67, DOI: 10.1007/978-3-319-03759-2_7, © Springer International Publishing Switzerland 2014

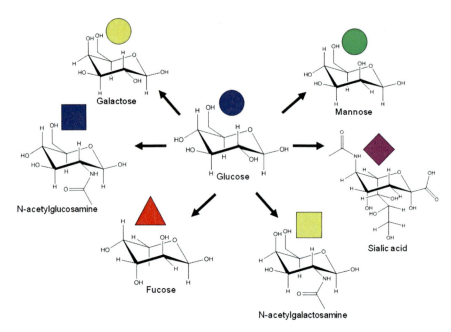

**Fig. 1** Structure and symbolic representations of common carbohydrates. Glucose is central to carbohydrate biosynthesis because it is made de novo from carbon dioxide and water during photosynthesis

would cite alternative splicing. We, too, believe that these phenomena contribute to biological complexity. Nevertheless we would argue that the greatest amplification of genomic information occurs after genes have been translated into proteins when the latter become modified by a myriad of functionalities. Moreover, one type of post-translational modification, namely glycosylation, results in the greatest diversity of the products of gene expression in all forms of life [1].

Today it has been well-established that protein N- and O-glycosylation (the covalent attachment of carbohydrate sequences to the side-chains of asparagine and serine or threonine, respectively) is a phenomenon shared by all domains of life. In addition, protein glycosylation has been demonstrated to be an essential requirement, rather than just an intriguing decoration. For example, correct glycosylation ensures that the plethora of proteins which eukaryotic cells use to transmit, receive and respond to chemical, electrical and mechanical signals, are expressed in functionally active forms in the right places. The information such glycoproteins mediate is essential for cells to pass through the different stages of development that occur in an organism [1, 2].

Carbohydrates have enough structural diversity to play a pivotal role as informational molecules on cell surfaces Fig. 1.

Importantly, they are in "the right place" to act as such. All eukaryotic cells are coated with a carbohydrate layer, referred to as the glycocalyx. It consists of glycoproteins and glycolipids embedded in the cell membrane, together with proteoglycans, another class of carbohydrate biopolymer, which may be loosely associated with the eukaryotic cell surface. Prokaryotes also express glycoproteins on their surfaces. Among the prokaryotic glycoproteins, the best understood are S-layers, pilins and flagellins, plus a selection of cell surface and secreted proteins which are known to be involved in adhesion and/or biofilm formation [3, 4]. Significantly, complex carbohydrates are often highly branched and each residue can be linked to another in any of several positions on each sugar ring. This allows the formation of a large number of oligosaccharide structures from a relatively small repertoire of building blocks. Indeed even greater diversity is often conferred by the addition of functional groups such as sulfates, phosphates, acetyl and methyl groups.

How do carbohydrates on cell surfaces fulfil their "information" roles? This is most often achieved by engagement with partner molecules on other cells thereby triggering adhesive and/or signalling events. These carbohydrate binding partners are called lectins [5]. Thanks largely to the Consortium for Functional Glycomics (CFG) (which was funded by the US National Institutes of Health to provide tools and resources to the international research community to understand the role of carbohydrate-protein interactions), scientists from all disciplines can readily access information pertaining to how surface carbohydrates and complementary lectins on opposing cell surfaces mediate cell-to-cell recognition. Thus the CFG website [2] provides a rich source of information and data which facilitates the engagement of researchers, unfamiliar with carbohydrates, with experts working in the field of glycobiology. Figure 2 illustrates several of the best understood biological interactions where carbohydrate-lectin recognition plays a central role. The meaning of the symbols used in the figure is explained in Fig. 2.

The tools of modern mass spectrometry have been crucial for unravelling the carbohydrate mediated processes exemplified in Fig. 2 [6–10]. Mass spectrometry is an enormously powerful tool for high sensitivity sequencing of complex carbohydrates. Its versatility permits the analysis of all families of glycopolymers. Moreover, complex mixtures of glycoproteins are not a problem for mass spectrometric analysis. Indeed, glycomic methodologies are capable of defining the carbohydrate sequences constituting the glycocalyx of tissues or cells without the need for time consuming purifications [11]. Glycomics research of the past decade, much of it supported by the CFG, has yielded substantial quantities of public data which are facilitating worldwide research addressing the roles of carbohydrates and lectins in complex systems [2, 12].

We hope that those reading this article are now stimulated to learn more about glycosylation and biological complexity. If this is the case, the CFG website is an excellent place to start your journey [2]. To whet your appetite we end our article with an introduction to an evolving story in glycobiology which has as its central character a famous carbohydrate moiety called sialyl Lewis$^x$ (Fig. 2).

First identified in rat brain glycoproteins in the 1970s, this carbohydrate was revealed, by the emerging glycomic strategies of the mid 1980s, to be present on human white blood cells and enriched in cancer cells. A few years later the Selectins were discovered. These constitute a lectin family that recognise sialyl Lewis$^x$ as their primary ligand. The Selectins play pivotal roles in lymphocyte trafficking and recruit-

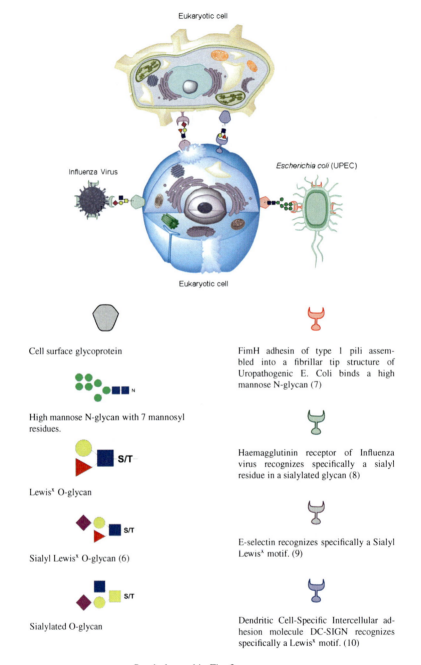

Symbols used in Fig. 2

**Fig. 2** Glycan–lectin recognition is key to cell-to-cell communication

ment of neutrophils to sites of inflammation. Their discovery energised the field of glycobiology and spawned numerous biotech companies intent on developing new anti-inflammatories and anti-cancer agents. The high hopes for glyco-therapeutics that prevailed in the early 1990s continue to this day, but are now tempered by realism i.e. it takes a very long time to understand the processes mediated by carbohydrate recognition and even longer to develop effective therapies based on intercepting these processes.

Very recently sialyl Lewis[x] has re-appeared in the headlines of both scientific and lay articles. This is because of exciting discoveries concerning human reproduction [13]. Ultra-high sensitivity mass spectrometric analyses have now provided the first molecular insights into the recognition processes occurring at the very start of human life, when a single sperm first engages with the surface of a human egg. This research has shown that multiple sialyl Lewis[x] sequences are attached to the proteins constituting the jelly-like coat of the human egg, which is called the zona pellucida. Remarkably the density of sialyl Lewis[x] moieties on the human egg is orders of magnitude higher than on white blood cells, consistent with it playing a pivotal role in sperm recognition [13]. Interestingly human sperm do not express any of the known Selectins. Hence the race is now on to find the putative "Selectin-like" molecule on sperm that binds to the sialyl Lewis[x] sequence on the human egg.

# References

1. A. Varki, R. Cummings, J. Esko, H. Freeze, P. Stanley, C. Bertozzi, G. Hart, M. Etzler, *Essentials of Glycobiology*, 2nd edn. (Cold Spring Harbor Laboratory, Cold Spring Harbor, New York, 2009)
2. Consortium for Functional Glycomics (CFG) web page, http://www.functionalglycomics.org/fg/
3. C. Szymanski, B. Wren, Protein glycosylation in bacterial mucosal pathogens. Nature Rev. Microbiol. **3**(3), 225–237 (2005)
4. A. Dell, A. Galadari, F. Sastre, P. Hitchen, Similarities and differences in the glycosylation mechanisms in prokaryotes and eukaryotes. Int. J. Microbiol. **2010**, 148178 (2010). doi:10.1155/2010/148178
5. N. Sharon, H. Lis, History of lectins: from hemagglutinins to biological recognition molecules. Glycobiology **14**(11), 53–62 (2004)
6. M. Fukuda, E. Spooncer, J. Oates, A. Dell, J. Klock, Structure of sialylated fucosyl lactosaminoglycan isolated from human granulocytes. J. Biol. Chem. **259**(17), 10925–10935 (1984)
7. C. Jones, J. Pinkner, R. Rotht, J. Heusert, A. Nicholes, S. Abraham, S. Hultgren, Fimh adhesin of type 1 pili is assembled into a fibrillar tip structure in the enterobacteriacea. Biochemistry. Proc. Natl. Acad. Sci. U.S.A. **92**, 2081–2085 (1995)
8. W. Weis, J. Brown, S. Cusack, J. Paulson, J. Skehel, D. Wiley, Structure of the influenza virus haemagglutinin complexed with its receptor, sialic acid. Nature **333**, 426–431 (1988)
9. M. Sperandio, C. Gleissner, K. Ley, Glycosylation in immune cell trafficking. Immunol. Rev. **230**(1), 97–113 (2009)
10. K. van Gisbergen, M. Sanchez-Hernandez, T. Geijtenbeek, Y. van Kooyk, Neutrophils mediate immune modulation of dendritic cells through glycosylation-dependent interactions between mac-1 and dc-sign. J. Exp. Med. **201**(8), 1281–1292 (2005)

11. S.J. North, J. Jang-Lee, R. Harrison, K. Canis, M. Nazri, A. Trollope, A. Antonopoulos, P.C. Pang, P. Grassi, S. Al-Chalabi, T. Etienne, A. Dell, S.M. Haslam, Chapter two-mass spectrometric analysis of mutant mice. Methods Enzymol. **478**, 27–77 (2010)
12. A. Antonopoulos, S.J. North, S.M. Haslam, A. Dell, Glycosylation of mouse and human immune cells: insights emerging from n-glycomics analyses. Biochem. Soc. Trans. **39**(5), 1334–1340 (2011)
13. P.C. Pang, P. Chiu, C.L. Lee, L.Y. Chang, M. Panico, H.R. Morris, S.M. Haslam, K.H. Khoo, G.F. Clark, W.S. Yeung, A. Dell, Human sperm binding is mediated by the sialyl-lewis$^x$ oligosaccharide on the zona pellucida. Science **333**(6050), 1761–1764 (2011)

# Some Thoughts on the Ontogenesis in B-Cell Immune Networks

**Elena Agliari, Adriano Barra, Silvio Franz and Thiago Pentado-Sabetta**

**Abstract** In this paper we focus on the antigen-independent maturation of B-cells and, via statistical mechanics tools, we study the emergence of self/non-self discrimination by mature B lymphocytes. We consider only B lymphocytes: despite this is an oversimplification, it may help to highlight the role of B-B interactions otherwise shadowed by other mechanisms due to helper T-cell signalling. Within a framework for B-cell interactions recently introduced, we impose that, during ontogenesis, those lymphocytes, which strongly react with a previously stored set of antigens assumed as "self", are killed. Hence, via numerical simulations we find that the resulting system of mature lymphocytes, i.e. those which have survived, shows anergy with respect to self-antigens, even in its mature life, that it to say, the learning process at ontogenesis develops a stable memory in the network. Moreover, when self-antigen are not assumed as purely random objects, which is a too strong simplification, but rather they are extracted from a biased probability distribution, mature lymphocytes displaying a higher weighted connectivity are also more affine with the set of self-antigens, ultimately conferring strong numerical evidence to the first postulate of autopoietic theories (e.g. Varela and Counthino approaches), according to which the most connected nodes in the idiotypic network are those self-directed.

---

E. Agliari
Dipartimento di Fisica, Sapienza Universitá di Roma, Parma, Italy
e-mail: elena.agliari@roma1.infn.it

E. Agliari
INFN, Gruppo Collegato di Parma, Parma, Italy

A. Barra (✉)
Dipartimento di Fisica, Sapienza Università di Roma, Roma, Italy
e-mail: adriano.barra@roma1.infn.it

S. Franz
Laboratoire de Physique Théorique et Modèles Statistiques, Universitè Paris-Sud 11, Orsay, France

T. Pentado-Sabetta
École Polytechnique, Paris, France

M. Delitala and G. Ajmone Marsan (eds.), *Managing Complexity, Reducing Perplexity*, Springer Proceedings in Mathematics & Statistics 67, DOI: 10.1007/978-3-319-03759-2_8,
© Springer International Publishing Switzerland 2014

**Keywords** Disordered statistical mechanics · Lymphocyte networks · Self/non-self discrimination

# 1 Introduction

Immunology is probably one of the fields of science which is experiencing the greatest amount of discoveries in these decades: As the number of experimental works increases, the need for minimal models able to offer a general framework where to properly locate experimental findings is a must for modelers interested in this field.

The purpose of the immune system is to detect and neutralize molecules, or cells (generically called antigens) potentially dangerous for the body, without damaging healthy cells [1]. The humoral response performed by B lymphocytes consists in analyzing the antigen, then the clone/s with the best matching antibody undergoes clonal expansion and releases specific immunoglobulins, which, in turn, are able to bind and neutralize pathogens. In order to achieve this goal, the immune system needs an enormous number of different clones, each having a particular receptor for antigens. As these receptors are generated at the genetic level randomly via somatic mutation, the body may produce lymphocytes attacking not only dangerous invaders (e.g. viruses), but also internal agents. The latter are referred to as self-reactive lymphocytes, which, if not carefully checked, may induce autoimmunity, an obviously unwanted feature.

In order to avoid auto-immunity, at least two mechanisms are thought to work: B-cells are generated, and maturate, in the bone marrow, where they are exposed to the so-called "negative selection rule".[1] More precisely, these lymphocytes are made to interact with an available repertoire of self-antigens, namely molecules/cells belonging to the host body, and those who are found to respond to them (so potential autoimmune B-cells) are induced to apoptosis, in such a way that only B-cells unable to attach to the available self survive[2] and share the freedom of exploring the body thereafter [1].

In fact, it is widely accepted that the bone marrow produces daily $\sim 10^7$ B cells, but only $\sim 10^6$ are allowed to circulate in the body, the remaining 90% undergo apoptosis since targeted as self-reactive [21]: as shown for instance by Nemazee and Burki [15], this depletion of the potential defense is due to the negative selection

---

[1] We only stress here that there exist strong differences between B-cell maturation in the bone marrow and T-cell maturation in the thymus [7, 12, 13]. Unlike TCR (T cell receptor), that evolved to recognize characteristic patterns of pathogens, BCR (B cell receptor) is primarily diversified in random fashion and has not evolved to recognize a particular structure. Therefore each B cell can not discriminate self versus non self alone [9].

[2] Strictly speaking, negative selection requires that newborn lymphocytes also display a non-null binding strength with at least a self-antigen, probably to avoid antibodies completely cut off from the host [10].

(clonal deletion) of immature B-cells expressing self-reactive antibodies or too low reactive ones.

As only a fraction of self-antigens are present into the bone marrow, self-reactive lymphocytes not expressing specific receptors (BCR) against the available self are allowed to circulate freely by this first security procedure. Hence, another mechanism must act at peripherals levels (i.e. in the lymphonodes, spleen and liver). Indeed, Goodnow was able to show experimentally [8] that these self-reactive lymphocytes actually exist in the body but, instead of undergoing apoptosis, they experience anergy in their responses, namely, under a proper stimulus, they do not respond. The main strand for explaining anergy and the consequent ability for self/non-self discrimination is via helper double signalling [9], nonetheless other mechanisms are expected to cooperate. Among these, collective features due to interactions among mature B cells may play a role and shall be investigated here trough statistical mechanics simulations.

The plan of the paper is as follows: in Sect. 2 we describe how the idiotypic network is generated and its main features; in Sect. 3 we develop the first approach to ontogenesis modeling, where we arbitrarily label as "self" a given amount of randomly generated antigens and check the subsequent growth of the network made of lymphocytes unable to attack these self-antigens. In Sect. 4 we develop an alternative approach, where we remove the (biological unreasonable) hypothesis that self-proteins are random objects and we deal with "correlated" self-antigens; impressively we find that in the correlated case the final repertoire not only correctly avoids to attach self, but also displays the peculiar topological structure suggested by Varela and coworkers, namely that nodes with high weighted connectivity are all self-directed. Finally, Sect. 5 contains discussions and comments on our results.

## 2 The Minimal Model

In this work we rely on the model introduced and developed in [2, 3], which achieves a description of the B-cell network able to recover as "emergent properties" basic facts such as low-dose tolerance, bell-shape response, memory features and self/non-self discrimination. However, within that framework the ability of the system to discriminate between self and non-self was recovered only at a cooperative level, in agreement with Varela and Coutinho [11, 18, 20]: clones which poorly interact with others are thought of as non-self-directed since they can easy respond to external fields (roughly speaking are more approximable as single particles), while clones which interact strongly and with a large number of other clones are thought of as self-directed since they experience a deep quiescent signal from nearest neighbors, which keeps them in a state of anergy. Here we want to move over and show that such mechanism regulation stems from and works synergically with negative selection.

Before proceeding, we briefly summarize the main features characterizing the interactions between B lymphocytes, ultimately leading to an idiotypic network; for more details we refer to [2, 3].

The system is made up of an ensemble of $N$ different clones, each composed of $M$ identical lymphocytes; a given lymphocyte $i$, is then described by the dichotomic variable $\sigma_i^\alpha = \pm 1$, with $\alpha = 1, ..., M$, and $i = 1, ..., N$, such that the value $-1$ denotes an anergic/absent state (low level of antibodies secretion) while the value $+1$ a firing state (high level of antibodies secretion). The antibodies secreted by a lymphocyte carry the very same idiotipicity expressed by the receptors of the secreting B cell. A generic antibody is represented by a binary string of length $L$, encoding the expression of $L$ epitopes.[3] In order to check immune responses we need to introduce the $N$ order parameters $m_i$ as local magnetizations:

$$m_i(t) = \frac{1}{M} \sum_{\alpha=1}^{M} \sigma_i^\alpha(t). \tag{1}$$

From the magnetizations $m_i \in [-1, 1]$, we can define the concentrations of the firing lymphocytes belonging to the $i$th family as $c_i(t) \equiv \exp[\tau(m_i(t) + 1)/2]$, where $\tau = \log M$, (see e.g. [4, 19]). Further, we introduce the Hamiltonian $H$ which encodes the interactions among lymphocytes as well as the interactions between lymphocytes and the external antigens:

$$H = H_1 + H_2 = -N^{-1} \sum_{i<j}^{N,N} J_{ij} m_i m_j - c \sum_{i}^{N} h_i^k m_i, \tag{2}$$

where $J_{ij}$ represents the coupling between clones $i$ and $j$, while $h_i^k$ represent the coupling between the clone $i$ and a given antigen $k$ (still represented by means of a binary string of length $L$) presented to the system and whose concentration is tuned by $c$.

The interaction matrix $\mathbf{J}$ and, similarly, the couplings $\mathbf{h}$, are built up as follows [2, 3]. Given two strings $\xi_i$ and $\xi_j$, representing the idiotipicity of two clones, their $\mu$-th entries are said to be complementary, iff $\xi_i^\mu \neq \xi_j^\mu$ so that the overall number of complementary entries $c_{ij} \in [0, L]$ can be written as $c_{ij} = \sum_{\mu=1}^{L} [\xi_i^\mu (1 - \xi_j^\mu) + \xi_j^\mu (1 - \xi_i^\mu)]$. Following biological arguments the affinity between two antibodies is expected to depend on how much complementary their structures are, hence, we introduce the functional

$$f_{\alpha,L}(\xi_i, \xi_j) \equiv [\alpha c_{ij} - (L - c_{ij})], \tag{3}$$

where $\alpha \in R^+$ quantifies the difference in the intensities of attractive and repulsive contributes. Notice that $f_{\alpha,L}(\xi_i, \xi_j) \in [-L, \alpha L]$ provides a measure of how "affine" $\xi_i$ and $\xi_j$ are. When the repulsive contribute prevails, the two antibodies do not match each other and the coupling among the corresponding lymphocytes $J_{ij}(\alpha, L)$

---

[3] The string length is assumed to be the same for any antibody following the fact that the molecular weight for any immunoglobulin is accurately close to $15 \times 10^4$ [5].

Some Thoughts on the Ontogenesis in B-Cell Immune Networks

is set equal to zero, conversely, we take $J_{ij}(\alpha, L) = \exp[f_{\alpha,L}(\xi_i, \xi_j)]/\langle \tilde{J} \rangle_{\alpha,L}$, being $\langle \tilde{J} \rangle_{\alpha,L}$ the proper normalizing factor so to keep unitary the average coupling.

As mentioned above, the generic antigen presented to the system can be modeled as well by means of a binary string $\xi_k$ and the rules determining the interaction strength between the $i$-th clone and the antigen are the same as for interaction between two antibodies, hence leading to the coupling $h_i^k$.

We finally recall that, from a statistical mechanics perspective, the Hamiltonian, calculated for a given configuration of magnetizations $\{m_i\}_{i=1,...,N}$, represents the "energy" pertaining to that configuration and, according to thermodynamic prescriptions, the system spontaneously tries to rearrange in order to minimize it. Since the coupling matrix is symmetric, it is possible to construct a dynamics satisfying the detailed balance and relaxing to Maxwell-Boltzmann distribution [19]. In the following this is realized using a standard Glauber single spin-flip dynamics.

We also stress that, as simulations with a realistic amount of clones are prohibitive in terms of CPU time, we worked at smaller repertoire sizes and tested the robustness of results trough finite-size-scaling analysis (see Fig. 1).

## 3 Random Ontogenesis

As we mentioned, during ontogenesis, those B-cells interacting strongly with self-antigens undergo negative selection and are deleted. Here we mimic this process by implementing the following learning rule: At the beginning, and once for all, $N_S$ vectors are randomly drawn from a uniform distribution and stored as "self". Then, we extract sequentially and randomly (again from a uniform distribution) new strings representing newborn lymphocytes and those which are able to bind strongly to at least one self-antigen from the set $N_S$ are killed, otherwise they are retained to build up the mature repertoire. The process is iterated until the size $N$ for the repertoire is reached.

The resulting system is therefore characterized by the parameters $N, N_S, L, \alpha$, where the interaction ratio $\alpha$ is kept fixed and equal to $\alpha = 0.7$ following biological evidences [2] and we also fix the scaling between $N$ and $L$ as $L = \gamma \log N$, according to bio-physical arguments [2]; here we choose $\gamma = 3$. As for $N_S$, we take it equal to a fraction of $N$. This allows to fulfill the relatively small survival probability for newborn B cells [21] and still retains a set of self-antigens vanishing with respect to the whole set of possible antigens, i.e. $N_S/2^L < \exp(-L(\log 2 - 1/\gamma))$. Of course, when $N_S = 0$ the original model [2, 3] is recovered. Finally, the binding between a newborn B and a self-antigen is considered to be strong if the number of complementarities between the related strings is larger than $3L/4$.

Once the repertoire has been created, external antigens are presented to it and responses are checked. First, we test its ability in self/non-self discrimination by presenting to the system a field composed only by self-antigens and measuring the resulting magnetization. Indeed, we find that anergy to self is completely fulfilled

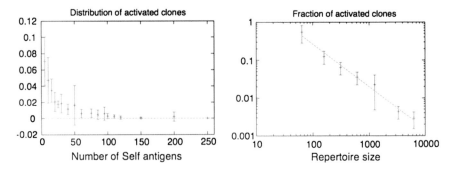

**Fig. 1** *Left* Distribution of the activated clones for an immune network at rest built up by $N = 628$ clones versus the amount of self-antigens used to generate the repertoire with antibodies made of by strings of $L = 11$ epitopes. *Right* Finite size scaling of the system. Averaged response of the network created trough a repertoire with $L = 8, \ldots, 14$ epitopes (keeping the fractions of the present clones and self-antigens constant) against one (randomly chosen) antigen of the repertoire itself. Coherently with the request that only a finite fraction of clones remains active increasing the network size, the fit is obtained trough $O(N^{-1})$ power (the fit with $N^x$ gives $x \sim -1.12$)

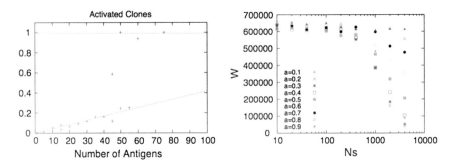

**Fig. 2** *Left* Fraction of the activated clones as a function of the antigens presented to the system: The system is made up of by 3352 clones and 50 self-antigens. *Right* Averaged weighted connectivity for different repertoires generated increasing the size of the experienced self $N_S$ at ontogenesis. For the latter various degree of correlation $a$ have been tested as explained by the legend. Here we fixed $N = 2N_s$ and $\gamma = 3$

(not shown in plots), for each experienced field made of by $1, \ldots, N_S$ self-antigens. Conversely, when antigens presented do not belong to self, the fraction of the activated clones grows as the number of antigens presented increases, as reported in Fig. 2 (left), eventually falling into a chronic activation state. This behavior can be easily understood from the perspective of spin glasses, due to the analogy between the system under study and a diluted random field model in the presence of a magnetic field: at low temperature, it undergoes a first order phase transition for a critical value of the external field [6, 14].

Furthermore, by enlarging the set of self-antigens $N_S$ (at fixed repertoire size $N$), the matching between a generic self-antigen and the mature repertoire gets sharper

# Some Thoughts on the Ontogenesis in B-Cell Immune Networks

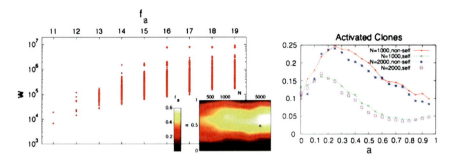

**Fig. 3** *Left* The figure envisages the correlation between the weighted degree $w$ of a node and its (maximum) affinity $f_a$ with the strings stored as self. Notice that larger values of $w$ correspond to larger values of $f_a$. Such correlation has been measured in terms of Spearman correlation coefficient $r_s$ which has been represented in the inset as a function of $a$ and $N$. The black star * corresponds to $N = 5000$ and $a = 0.45$, which are the parameters used for the main plot. *Right* Fraction of activated clones as a function of $a$ and for different sizes of the repertoire, as explained by the legend. The response of the system is measured in the case a self-antigen (*dotted lines*) and a external, non-self antigen (*continuous line*) is presented

so that only a small amount of highly affine clones is able to respond (see Fig. 1, right).

## 4 Correlated Ontogenesis

Despite a certain degree of stochasticity seems to be present even in biological systems [1], self-proteins are not completely random objects [16, 17]. In order to account for this feature we now generate the repertoire of self-antigens according to a probability distribution able to induce a correlation between epitopes of self-antigens. Seeking for simplicity we adopt the following

$$P_{self}(\xi_i^\mu = +1) = (1+a)/2, \ P_{self}(\xi_i^\mu = 0) = (1-a)/2,$$

where $a$ is a parameter tuning the degree of correlation: when $a = 0$ we recover the unbiased situation described in the previous section, while increasing (decreasing) $a \to +1$ ($a \to -1$) we move towards stronger correlation.

Figure 3 shows that the correlation between the weighted degree of a node and its affinity with self is numerically confirmed, and turns out to be larger for intermediate values of $a$. As a result, the system is expected to respond more strongly to non-self antigens, consistently with an healthy behavior (see Fig. 3, right and Fig. 4, left).

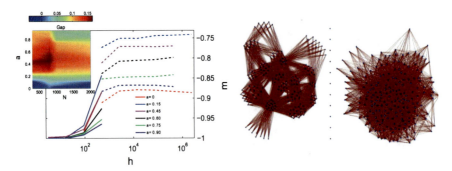

**Fig. 4** *Left* the main plot shows the local immune response, in terms of clonal magnetization $m$, versus the resulting field $h$ due to the presentation of an antigen, either self (*continuous line*) and non-self (*dotted line*). The clone considered for the measure of the system response is the one with larger affinity with the string presented. Here we fixed $N = 5000$ and $\gamma = 3$. Moreover, several degrees of correlation $a$ for the stored self are considered, as shown by the legend. For any of them there exists a value $h^*$, which typically works as upper bound for fields generated by self agents and as lower bound for fields generated by non-self agents. Interestingly, at $h^*$ the resulting magnetizations exhibit a gap $m_{\text{non-self}}(h^*) - m_{\text{self}}(h^*)$, which is shown in the inset as a function of $a$ and $N$. *Right* Examples of the resulting idiotypic network with $a = 0.25$, $N = 200$ and $N_S = N$ (*right*) and $N_S = 2N$ (*left*)

## 5 Discussion

In this paper we investigated the effects of negative selection occurring during the ontogenesis of B-cells. First we showed the ability of the system to develop memory of the self experienced at ontogenesis, in such a way that cells self-directed behave anergetically even in the mature repertoire. We also get a numerical confirmation of Varela's suggestions [18, 20], according to which nodes with high (weighted) connectivity can be looked at as "self-directed". Therefore, ontogenesis acts as a learning process that, from one side, teaches to each single lymphocyte not to attack the proteins seen during maturation, and on the other side induces a correlation among idiotipicity yielding a possible regulatory role for the mature B-cell network. In this way, Varela's assumption is moved from a postulate to a physical consequence of a correlated learning process.

In this process a fundamental requisite is that self-proteins are not purely random object, but they share a certain degree of correlation. Here we introduce this bias in the simplest way just to show the idea; more biological patterns can possibly be implemented.

Future development should include T-helper interactions as well as an exploration of the relation between the amount of stored self-antigens in ontogenesis and the stability of the mature response against the number of encountered pathogens.

This research belongs to the strategy of exploration funded by the MIUR trough the FIRB project *RBFR08EKEV* which is acknowledged.

# References

1. A.K. Abbas, A.H. Lichtman, J.S. Pober, *Cellular and Molecular Immunology* (Elsevier, Amsterdam, 2007)
2. A. Barra, E. Agliari, J. Stat. Mech. P07004 (2010)
3. A. Barra, E. Agliari, Physica A **389**, 24 (2010)
4. A. Di Biasio, E. Agliari, A. Barra, R. Burioni, Theoret. Chem. Acc. in press (2011)
5. W.J. Dreyer, J.C. Bennett, Nobel Prize Lecture (1965)
6. S. Franz, G. Parisi, Phys. Rev. Lett. **79**, 2486 (1997)
7. R.A. Goldsby, T.J. Kindt, B.A. Osborne, *Kuby: Immunology* (UTET Press, New York, 2000)
8. C.C. Goodnow et al., Nature **334**, 676–682 (1988)
9. D. Kitamura, *How the Immune System Recognizes Self and Nonself* (Springer, Shinano, 2008)
10. A. Kosmrlj et al., Phys. Rev. Lett. **103**, 068103 (2009)
11. I. Lundkvist, A. Coutinho, F. Varela, D. Holmberg, Proc. Natl. Acad. Sci. **86**(13), 5074–5078, (1989).
12. R. Mehr, A. Globerson, A.S. Perelson, J. Theor. Biol. **175**, 103–126 (1995)
13. S.J. Merrill, R.J. De Boer, A.S. Perelson, Rocky Mountain J. Math. **24**, 213–231 (1994)
14. M. Mezard, G. Parisi, M.A. Virasoro, *Spin Glass Theory and Beyond* (World Scientific, Singapore, 1988)
15. D.A. Nemazee, K. Burki, Nature **337**, 562 (1989)
16. A.S. Perelson, *Theoretical Immunology* (Addison-Wiley Publ, Santa Fe Institute, 1988)
17. S. Rabello, A.C.C. Coolen, C.J. Pérez-Vicente, F. Fraternali, J. Phys. A **41**, 285004 (2008)
18. J. Stewart, F.J. Varela, A. Coutinho, J. Autoimmun. **2**, 1523 (1989)
19. C.J. Thompson, *Mathematical Statistical Mechanics* (MacMillan Company, New York, 1982)
20. F.J. Varela, A. Countinho, Immun. Today **12**(5), 159 (1991)
21. Wardemann et al., Science **301**, 1374–1377 (2003)

# Mathematical Modeling of Cancer Cells Evolution Under Targeted Chemotherapies

**Marcello Delitala and Tommaso Lorenzi**

**Abstract** This chapter focuses on selection and resistance to drugs in an integro-differential model describing the dynamics of a cancer cell population exposed to targeted chemotherapies. Mutations, proliferation and competition for resources are assumed to occur under the cytotoxic action of targeted therapeutic agents. The results obtained support the idea that cancer progression selects for highly proliferative clones. Moreover, it is highlighted how targeted chemotherapies might act as an additional selective pressure leading to the selection for the fittest, and thus eventually most resistant, cancer clones.

**Keywords** Structured populations · Cancer modeling and therapy Evolution · Concentration phenomena · Integro-differential equations

---

M. Delitala (✉) · T. Lorenzi
Department of Mathematical Sciences, Politecnico di Torino,
Corso Duca degli Abruzzi 24, 10129 Torino, Italy
e-mail: marcello.delitala@polito.it

T. Lorenzi
Laboratoire Jacques-Louis Lions, Sorbonne Universités, UPMC Univ Paris 06, UMR 7598, Paris, France
e-mail: tommaso.lorenzi@upmc.fr

T. Lorenzi
CNRS, Laboratoire Jacques-Louis Lions, UMR 7598, F-75005 Paris, France

T. Lorenzi
EPC MAMBA, INRIA-Paris-Rocquencourt, Domaine de Voluceau, BP105, Le Chesnay Cedex 78153, France

M. Delitala and G. Ajmone Marsan (eds.), *Managing Complexity, Reducing Perplexity*,
Springer Proceedings in Mathematics & Statistics 67, DOI: 10.1007/978-3-319-03759-2_9,
© Springer International Publishing Switzerland 2014

# 1 Introduction

Solid tumors can be seen as heterogeneous aggregates composed of cells carrying different mutations, which compete for space and resources (e.g. oxygen and glucose) and try to evade the predation exerted by the immune system and by therapeutic agents [9].

The fitness of neoplastic clones (i.e. their ability to survive and reproduce) is shaped by different selective pressures, which can vary from one organ to another [7]. This implies that, depending on the environmental context, the same mutation can be advantageous/deleterious/neutral (i.e. it can increase/decrease/not affect the cellular fitness).

Tumor evolution usually privileges the selection of cells endowed by mutations with high proliferative abilities. Even more, the exposure to chemotherapeutic drugs may reinforce the selection for the fittest, and thus eventually most resistant, cancer clones. As a result, as time goes by, a rapid evolution toward highly malignant genotypic-phenotypic profiles can occur within tumor aggregates, which is likely to be the main reason why targeted chemotherapeutic treatments may fail in curing cancer.

An integro-differential model for the dynamics of cancer cells is proposed here, which aims at highlighting those phenomena that play a key role in tumorigenesis, focusing on the aspects related to tumor progression, intra-tumor heterogeneity and response to targeted cytotoxic therapies. Such model can be viewed as a simplified version of that developed in [2] and relies on the mathematical structures proposed in [1, 10]. In more detail, the contents of this chapter are organized as follows:

Section 2 is devoted to outline the essential features of the biological phenomena under consideration and to present the mathematical model.

Section 3 is meant to summarize the emerging phenomena highlighted by the model. The results of numerical simulations are reported and related biological interpretations are provided.

# 2 The Model

In this chapter, we focus on a sample of cancer cells characterized by heterogeneous genotypic-phenotypic profiles (i.e. different expression levels of the genes involved in cancer progression and the related observable traits), exposed to Targeted Chemotherapeutic Agents (TCAs, in the sequel).

Moving toward a mathematical formalization, we look at the sample as a population structured by a continuous variable $u \in U := [0, 1]$, standing for the genotypic-phenotypic profile of the cells. Since we are interested, at this stage, in evolutionary aspects, phenomena involving geometrical and mechanical variables are not under consideration. The cell population is characterized by the function

$$f = f(t, u) : [0, T] \times U \to \mathbb{R}^+,$$

where the time variable $t$ is normalized with respect to the average life-cycle duration of cancer cells and parameter $T$ models the end time of observations. At any fixed time $t$, the quantity $f(t, u) \, du$ stands for the number of cells whose genotypic-phenotypic profile belongs to the volume element $du$ centered at $u$, normalized with respect to the total number of cells inside the system at time $t = 0$.

Macroscopic gross variables can be computed through integration. In particular, the total density of the population at time $t$ is defined as:

$$n(t) := \int_U f(t, u) du. \tag{1}$$

Cancer cells are exposed to the action of TCAs, considered as an additional population structured by a continuous variable $v \in V := U$, which is related to the genotypic-phenotypic profile of the cells that can be mainly recognized and attacked by the curing agents. This additional population is characterized by the function

$$g = g(t, v) : [0, T] \times V \to \mathbb{R}^+;$$

considerations analogous to the ones drawn about function $f$ hold for function $g$, as well.

The biological phenomena of interest are modeled according to the assumptions and the strategies below summarized. Mathematical details are close to the ones that we have previously introduced in [2]. In fact, compared to the model developed there, the one presented here can be seen, at least to a certain extent, as an essential and simplified version, which is nevertheless able to catch some interesting emerging behaviors.


## 2.1 Cell Mutations and Renewal

Net of cell renewal, mutations lead parent cells to generate daughter cells characterized by different genotypic-phenotypic profiles. Since mutations usually lead to small variations, we make use of a small parameter $\varepsilon$, measuring the average size of such changes, and a parameter $\alpha$, modeling the average probability for genotypic-phenotypic modifications, to define a mutation kernel $\mathcal{M}(u, u_*; \varepsilon)$ as follows:

$$\mathcal{M}(u, u_*; \varepsilon) := \begin{cases} \alpha\delta(u - (u_* \pm \varepsilon)) + (1 - 2\alpha)\delta(u - u_*), & \text{if } \varepsilon < u < 1 - \varepsilon \\ \alpha\delta(u - (u_* - \varepsilon)) + (1 - \alpha)\delta(u - u_*), & \text{if } 0 \leq u \leq \varepsilon \\ \alpha\delta(u - (u_* + \varepsilon)) + (1 - \alpha)\delta(u - u_*), & \text{if } 1 - \varepsilon \leq u \leq 1, \end{cases}$$

where $\delta$ is the Dirac's delta distribution. It is worth noting that kernel $\mathcal{M}$ has the structure of a probability density.

## 2.2 Cell Proliferation

In order to mimic the effects of cancer growth, we introduce a positive function $\kappa(u)$, which models the rate of cell proliferation net of apoptosis and it is assumed to be sufficiently smooth as well as to have a maximum value $\kappa^C$. Cells proliferate at different rates depending on the shape of function $\kappa$ and the clones expressing the most proliferating genotypic-phenotypic profiles duplicate at rate $\kappa^C$. It should be noted that we are modeling mutations separately from proliferation. In fact, these phenomena usually occur on different time scales and distinct parameters are required to model the related frequencies.

## 2.3 Competition for Resources Between Cancer Cells

Cellular proliferation is hampered by the competition for resources. Therefore, we assume that interactions can lead cancer cells to die at a rate defined by a positive smooth function $\mu(u)$, whose maximum value is identified by parameter $\mu^C$. Cells die, due to lack of resources, at different rates depending on the shape of function $\mu$. The clones dying at rate $\mu^C$ are characterized by those genotypic-phenotypic profiles endowing them with the weakest competitive abilities.

## 2.4 Competition Between Cancer Cells and Targeted Chemotherapeutic Agents

Targeted chemotherapies are able to selectively kill cancer cells characterized by specific genotypic-phenotypic profiles. The average effectiveness of TCAs is modeled by a parameter $\mu^T$, while a parameter $\theta^T$ is introduced as an average measure of the cancer-therapy interaction selectivity.

To sum up, a 6 parameters model is defined, where all the parameters are positive real numbers characterized by a well defined biological meaning:

$$\frac{\partial}{\partial t} f(t, u) = \underbrace{\int_U \mathcal{M}(u, u_*; \varepsilon) f(t, u_*) du_* - f(t, u)}_{\text{mutations and renewal}} + \underbrace{\kappa(u) f(t, u)}_{\text{cell proliferation}}$$

$$- \underbrace{\mu(u) f(t, u) \mathrm{n}(t)}_{\text{cell-cell competition}} - \underbrace{\mu^T f(t, u) \int_V e^{-\theta^T (v^* - u)^2} g(t, v^*) dv^*}_{\text{destruction due to TCAs}}, \quad (2)$$

which describes the net inlet of cells through the volume element $du$ centered at $u$ at time $t$.

Mathematical Modeling of Cancer Cells Evolution

With reference to TCAs, a detailed balance equation is not introduced to describe the dynamics of $g(t, v)$, which is supposed to be a given smooth function of its argument.

## 3 Computational Results and Critical Analysis

This section summarizes numerical results obtained by solving two distinct initial value problems linked to Eq. (2). Focusing on emerging behaviors, computational analysis are addressed to study how the dynamics of $f(t, u)$ is affected by the values of some critical parameters, which are selected case by case with explorative aims. In particular, simulations are meant to:

- enlighten the role played by the biological phenomena under consideration within dynamics of cancer cells, with particular reference to progression and heterogeneity aspects;
- reproduce the emergence of resistance to anti-cancer therapies and highlight the controversial role that targeted chemotherapies can play in cancer development.

With this aim, given the expressions of the parameter functions, we numerically solve the mathematical problems defined by endowing Eq. (2) with two different definitions of functions $g$, corresponding to the case where TCAs are not inoculated in the system or to the case where they are administered starting from time $t = T/2$, as well as with a given initial condition $f(t = 0, u)$, which mimics a sample where cells mainly express the genotypic-phenotypic profiles corresponding to $u = 0.15$ and $u = 0.85$, at the beginning of observations. Standard fixed point arguments can be used to prove that the Cauchy Problems here considered are well-posed in the sense of Hadamard (i.e. the solution exists, it is unique and depends continuously on the initial data).

The following considerations and assumptions hold along all simulations:

- We let genotypic-phenotypic changes to be small; thus, we set $\varepsilon = 0.001$.
- We arbitrarily define the proliferation rate $\kappa(u)$ in such a way that the most proliferating cells are characterized by four among the possible genotypic-phenotypic profiles, i.e.

$$\kappa(u) = \kappa^C \quad \Longleftrightarrow \quad u \in \{0.05, 0.25, 0.75, 0.95\}.$$

- Since the most proliferating cancer cells need more resources to survive than the others, we assume these cell to be prone to fail in the competition for resources. For this reason, we let function $\mu$ to be proportional to function $\kappa$, i.e.

$$\mu(u) = \beta\kappa(u), \quad 0 < \beta < 1.$$

- Function $g$ is defined in such a way that inoculated TCAs are mainly able to act against those cancer cells that express the genotypic-phenotypic profiles corresponding to $u = 0.25$ and $u = 0.75$.
- Parameters $\alpha$, $\mu^T$ and $\theta^T$ are set equal to suitable non-zero values selected with exploratory aim and we fix $T = 100$.

## 3.1 Effects of Mutations and Proliferation in Absence of Therapeutic Agents

We focus on the role played by mutation and proliferation phenomena in cancer dynamics, considering the additional definition for $g(t, v)$ that mimics a scenario where TCAs are not inoculated. The results summarized by the left panel of Fig. 1 illustrate how, when $\varepsilon$ is small, $f(t, u)$ concentrates, across time, around the points where $\kappa(u)$ attains its maximum, i.e. $u = 0.05$, $u = 0.25$, $u = 0.75$ and $u = 0.95$.

Since intra-tumor heterogeneity is due to the presence, within the same tumor aggregate, of cells expressing several genotypic-phenotypic profiles, these results support the idea that a strong reduction in heterogeneity occurs, if mutations cause small changes in the genotypic-phenotypic profiles. In fact, only cells endowed with strong proliferative abilities can survive inside the sample, while weakly proliferative mutants die out. Thus, we are led to the same conclusions drawn in [7]: cancer progression selects for highly proliferative clones.

From an evolutionary perspective, the left panel of Fig. 1 also highlights how branching patterns may arise in cancer dynamics, at least in those cases where, at the beginning of observations, tumor cells mainly express some genotypic-phenotypic profiles that are different from the most proliferating ones.

## 3.2 Controversial Role of Targeted Chemotherapeutic Agents

The following computational analysis are meant to deepen the role that TCAs can play in cancer evolution. In particular, we make a comparison between the numerical solutions obtained with the two expressions of $g(t, v)$ that mimic the scenarios with and without therapies.

Figure 2 shows how, if TCAs are inoculated in the system, the picks of $f(t, u)$ centered in $u = 0.25$ and $u = 0.75$ vanish over the time interval $(T/2, T]$, since, in the case at hand, $g(t, v)$ is assumed to be mainly concentrated in these points. Moreover, the same figure highlights how $f(T, 0.05)$ and $f(T, 0.95)$ are greater in the case with TCAs (solid lines) rather than in the case without therapeutic agents (dashed lines).

The results summarized by Fig. 2 support the idea that, in those cases where environmental conditions select for strong proliferative abilities and several sub-

# Mathematical Modeling of Cancer Cells Evolution

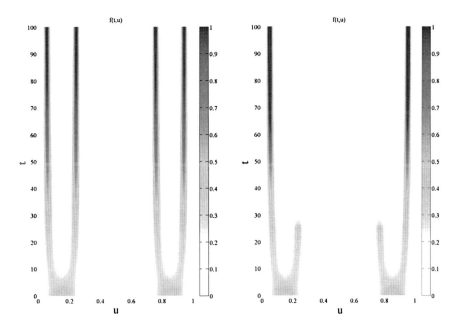

**Fig. 1** Dynamics of $f(t, u)$ in a small mutation regime (i.e. $\varepsilon \to 0$). *Left panel* in absence of therapeutic agents. Function $f(t, u)$ concentrates, across time, around the points where $\kappa(u)$ (i.e. the proliferation rate of cells) attains its maximum. This result supports the idea that, in the limit of small mutations, only cells endowed with strong proliferative abilities can survive inside the sample. *Right panel* in presence of targeted chemotherapeutic agents delivered at $t = 25$. Function $f(t, u)$ concentrates around the points where $\kappa(u)$ attains its maximum (i.e. $u = 0.05$, $u = 0.25$, $u = 0.75$ and $u = 0.95$) over the time interval $[0, 50]$, since we are letting $\varepsilon \to 0$. However, due to the fact that $g(t, v)$ is mainly concentrated around point $v = 0.25$ and $v = 0.75$, the picks of $f(t, u)$ centered in $u = 0.25$ and $u = 0.75$ vanish over the time interval $(50, 100]$

populations of highly proliferative clones are found inside the system, if TCAs cause the extinction of some sub-populations, the clonal expansion of cells in the other ones is intensified.

These simulations should be interpreted as a virtual version of some classic early experiments in evolutionary biology, which have been devoted to test whether the exposure of a sample population to a selective force causes new mutations to occur or selects for pre-existing mutants. The obtained results support the second case and reinforce the considerations drawn in [3, 6–8], suggesting that TCAs may introduce an additional selective pressure that reinforces the selection for the fittest clones. This is a well known outcome of many pharmacotherapies that, unless they full eradicate the mutated cells, generally fail in cancer treatment.

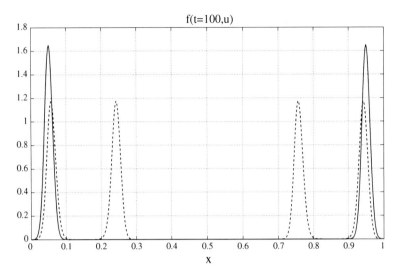

**Fig. 2** Comparison between the trend of $f(t = 100, u)$ with no therapies (*solid lines*) and with therapies (*dashed lines*), in a small mutation regime (i.e. $\varepsilon \to 0$). Targeted chemotherapeutic agents can select for the fittest genotypic-phenotypic profiles. In fact, if environmental conditions favor the selection of strong proliferative abilities in such a way that multiple sub-populations of highly proliferative clones are found inside the system (*dashed lines*), the inoculation of targeted chemotherapies can cause the extinction of one or more sub-populations so that the clonal expansion of cells in the other sub-populations is intensified (*solid lines*)

## 3.3 Critical Analysis

In this chapter we have proposed a simple mathematical model for the dynamics of cancer cells exposed to targeted chemotherapeutic treatments, with the aim of enlightening the causes for some emerging phenomena observed in tumor progression.

At first, we have analyzed the role that proliferation and mutations play in carcinogenesis, under a regime of small genotypic-phenotypic changes. Then, still assuming mutations to be small, we have examined the effects of targeted chemotherapeutic agents.

The obtained results suggest that cancer progression selects for strong proliferative cells, while targeted chemotherapies might act as an additional selective pressure, leading to the selection for the fittest, and thus eventually most resistant, cancer clones.

The same conclusions are obtained by other authors [4, 11] and support the development of the so-called adaptive therapies, which are principally aimed at a stabilization of tumor cells, instead of an unlikely full eradication [5].

**Acknowledgments** Partially supported by the FIRB project - RBID08PP3J.

# References

1. N. Bellomo, M. Delitala, From the mathematical kinetic, and stochastic game theory to modeling mutations, onset, progression and immune competition of cancer cells. Phys. Life Rev. **5**, 183–206 (2008)
2. M. Delitala, T. Lorenzi, A mathematical model for the dynamics of cancer hepatocytes under therapeutic actions. J. Theoret. Biol. **297**, 88–102 (2012)
3. J. Foo, F. Michor, Evolution of resistance to anti-cancer therapy during general dosing schedules. J. Theor. Biol. **263**, 179–188 (2010)
4. R.A. Gatenby, A change of strategy in the war on cancer. Nature **459**, 508–509 (2009)
5. R.A. Gatenby, A.S. Silva, R.J. Gillies, B.R. Frieden, Adaptive therapy. Cancer Res. **69**, 4894–4903 (2009)
6. N.L. Komarova, D. Wodarz, Drug resistance in cancer: principles of emergence and prevention. Proc. Natl. Acad. Sci. USA **102**, 9714–9719 (2005)
7. L.M. Merlo, J.W. Pepper, B.J. Reid, C.C. Maley, Cancer as an evolutionary and ecological process. Nat. Rev. Cancer **6**, 924–935 (2006)
8. F. Michor, M.A. Nowak, Y. Iwasa, Evolution of resistance to cancer therapy. Curr. Pharm. Des. **12**, 261–271 (2006)
9. P.C. Nowell, The clonal evolution of tumor cell populations. Science **194**, 23–28 (1976)
10. B. Perthame, *Transport Equations in Biology* (Birkhäuser, Basel, 2007)
11. A.S. Silva, R.A. Gatenby, A theoretical quantitative model for evolution of cancer chemotherapy resistance. Biol. Direct **5**, 25 (2010)

# Traveling Waves Emerging in a Diffusive Moving Filament System

**Heinrich Freistühler, Jan Fuhrmann and Angela Stevens**

**Abstract** Starting from a minimal model for the actin cytoskeleton of motile cells we derive a spatially one dimensional model describing populations of right and left moving filaments with intrinsic velocity, diffusion and mutual alignment. For this model we derive traveling wave solutions whose speed and shape depend on the model parameters and the type of alignment. We discuss possible wave profiles obtained from analytical investigations as well as waves emerging in numerical simulations. In particular, we will explicitly comment on the observed wave speeds and how they are related to the model parameters. Moreover, some particularly interesting patterns being composed of several wave profiles are discussed in some detail. Finally, we shall try to draw some conclusions for the full cytoskeleton model our system had emerged from.

**Keywords** Cytoskeleton · Nonlinear waves · Reaction diffusion advection equations

## 1 Motivation and Derivation of the Model

Understanding the mechanisms of actin driven cell motility is of great importance for a variety of biological processes such as wound healing, metastasis of cancer, immune response, and many others. In [3], a minimal, spatially one dimensional

---

H. Freistühler
Department of Mathematics and Statistics, University of Konstanz,
Fach 199, 78457 Konstanz, Germany

J. Fuhrmann (✉)
Institute of Mathematics, Johannes Gutenburg University Mainz,
Staudingerweg 9, 55123 Mainz, Germany
e-mail: fuhrmann@uni-mainz.de

A. Stevens
Institute for Computational and Applied Mathematics, University of Münster,
Einsteinstraße 62, 48149 Münster, Germany

M. Delitala and G. Ajmone Marsan (eds.), *Managing Complexity, Reducing Perplexity*,
Springer Proceedings in Mathematics & Statistics 67, DOI: 10.1007/978-3-319-03759-2_10,
© Springer International Publishing Switzerland 2014

model for the actin cytoskeleton of a potentially motile cell at rest was proposed in order to understand the polarization of the cytoskeleton upon some external stimulus which might drive the resting cell into directed motion.

We deduced a system consisting of four hyperbolic conservation equations for the densities of barbed ($B$) and pointed ($P$) ends of right (subscript $r$) and left (subscript $l$) actin filaments, respectively,

$$\partial_t B_r(t, x) + \partial_x (v_B(t, x, a) B_r(t, x)) = 0, \tag{1}$$
$$\partial_t B_l(t, x) - \partial_x (v_B(t, x, a) B_l(t, x)) = 0, \tag{2}$$
$$\partial_t P_r(t, x) + \partial_x (v_P(t, x, a) P_r(t, x)) = 0, \tag{3}$$
$$\partial_t P_l(t, x) - \partial_x (v_P(t, x, a) P_l(t, x)) = 0, \tag{4}$$

and a parabolic reaction diffusion equation for the actin monomer concentration

$$\partial_t a(t, x) - D \partial_{xx} a(t, x) = R(t, x, a, B_r, P_r, B_l, P_l). \tag{5}$$

Each of the hyperbolic equations is coupled to the parabolic equation by the dependence of the flux velocities $v_B$ and $v_P$ on the monomer density $a$. The reaction diffusion equation in turn receives input from the conservation laws via the reaction term describing the binding or release of monomers at polymerizing or depolymerizing filament ends.

Since in vivo, the cytoskeleton is permanently remodeled and filament tips are subject to thermal fluctuations we now want to investigate the behavior of the hyperbolic part upon additional effects like diffusion and mutual alignment of filaments while the monomer concentration is fixed to some specific value. Only upon including these effects we can step beyond the very initial steps of cell polarization and ask for aligned structures like stress fibers or lamellipodia.

To this end, let us assume constant parameters in the above model and a fixed monomer density $\bar{a}$ such that

$$v_B(t, x, \bar{a}) = v_P(t, x, \bar{a}) \equiv \bar{v}. \tag{6}$$

Moreover, we assume the filaments to be very short so that as an approximation we can identify barbed and pointed ends of either orientation, right and left.

We end up with two densities, $u_r$ and $u_l$, of filaments moving to the left or right, respectively, at velocity $\bar{v}$ which shall be put to one for simplicity. Their movement is now governed by the particularly simple system

$$\partial_t u_r + \partial_x u_r = \varepsilon \partial_{xx} u_r \tag{7}$$
$$\partial_t u_l - \partial_x u_l = \varepsilon \partial_{xx} u_l \tag{8}$$

where we also allowed for the diffusion of filaments at a diffusion coefficient $\varepsilon$. This can be interpreted as fluctuations of the rather small filaments in the crowded environment inside a cell full of proteins and other obstacles. As $\varepsilon$ takes small values,

Traveling Waves Emerging in a Diffusive Moving Filament System

this parabolic system can be understood as a slightly but singularly perturbed version of its hyperbolic limit which is obtained for $\varepsilon = 0$ and corresponds to the original hyperbolic part.

Introducing the total density $u = u_r + u_l$ and the polarization $w = u_r - u_l$ and allowing in addition for mutual alignment of the particles we can rewrite this into

$$\partial_t u + \partial_x w = \varepsilon \partial_{xx} u \tag{9}$$

$$\partial_t w + \partial_x u = \varepsilon \partial_{xx} w + f(u, w). \tag{10}$$

Here, the alignment term $f$ describes the ability of particles moving in one direction to reverse those with opposite direction of motion by mutual alignment.

Two families of alignment terms are considered which we will refer to as sublinear and superlinear type. For a given total density, say $u = 1$ for simplicity, the prototypical examples take the forms

$$f(1, w) = \alpha w \left(1 - w^2\right) \qquad \text{(sublinear)} \tag{11}$$

and

$$f(1, w) = \alpha w \left(1 + v\, w^2 - (v + 1)w^4\right) \qquad \text{(superlinear).} \tag{12}$$

## 2 Traveling Waves

In this section, we are going to deduce the existence of traveling wave solutions to the system (9), (10) and discuss some of their properties. In physical terms, these wave patterns correspond to fronts of filaments moving at constant velocity as they are observed in lamellae of moving cells (cf. [6]). Particular emphasis will be put on the possible wave velocities.

### 2.1 Traveling Waves for Two Reduced Problems

Before investigating the full problem we focus on a simple auxiliary problem. Let us consider the hyperbolic equations

$$\partial_t u + \partial_x w = 0 \tag{13}$$

$$\partial_t w + \partial_x u = f(u, w) \tag{14}$$

which result from (9) and (10) upon formally sending $\varepsilon$ to zero.

We are looking for solutions of the type $u(t, x) = U(x - ct)$ and $w(t, x) = W(x - ct)$ with the wave profiles $U$ and $W$ and the constant wave velocity $c$. The corresponding system of ordinary differential equations reads

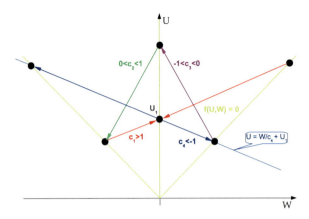

**Fig. 1** Possible heteroclinic orbits for system (15), (16) in the $W$–$U$ phase space. *Yellow lines* denote the equilibria of the system, *colored arrows* represent possible orbits corresponding to different velocities $c$

$$-cU' + W' = 0 \qquad (15)$$
$$-cW' + U' = f(U, W). \qquad (16)$$

As possible equilibria we identify the zeroes of $f$ which are

1. the non-polarized state $W = 0$ with constant total density $U > 0$ with equally many right and left moving filaments corresponding to a non-polarized cell,
2. and the totally aligned states $W = \pm U$ corresponding to all filaments moving in the same direction.

We only obtain wave profiles connecting any of the totally aligned state to the non-polarized state or vice versa. The corresponding wave velocities $c$ are only restricted by the conditions $c \neq 0$ and $c \neq \pm 1$. Possible examples of such profiles appearing as heteroclinic orbits in phase space are depicted in Fig. 1.

Passing from the hyperbolic problem (13), (14) to the full problem (9), (10) we briefly consider the intermediate problem (13), (10) with a diffusion term only in the equation for the polarization $w$. The system of traveling wave equations for this problem reads

$$-cU' + W' = 0 \qquad (17)$$
$$-cW' + U' = \varepsilon W'' + f(U, W). \qquad (18)$$

This can be reduced to a single second order equation for $W$ which in turn can be written as a first order system consisting of two equations. We are thus dealing with an effectively two dimensional phase space and can again find heteroclinic orbits between the equilibria which translate into traveling waves for the corresponding system of partial differential equations.

# Traveling Waves Emerging in a Diffusive Moving Filament System

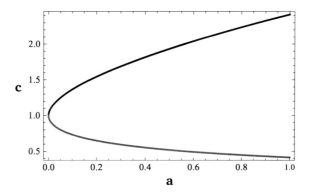

**Fig. 2** Precise dependence of the critical velocities $c^*$ (*upper branch*) and $c_*$ (*lower branch*) on the value $a := F_w \varepsilon$

Using a sublinear alignment term $f$, we find two critical wave velocities $c^* > 1$ and $c_* < 1$, determined by the parameters, which define the boundary between the existence and non existence of monotone traveling wave profiles connecting one of the fully polarized states $W = \pm U$ with the symmetric state ($W = 0$, $U = \bar{U}$), similar to the minimal wave velocity for the Fisher-KPP equation (cf. [5]).

More precisely, there are monotone waves between a totally aligned and the symmetric state exactly for velocities $c$ satisfying

$$0 < |c| \leq c_* \quad \text{or} \quad |c| \geq c^*. \tag{19}$$

Denoting by

$$F_w := \partial_w f(\bar{U}, 0)$$

the partial derivative of the alignment term with respect to the polarization, evaluated at the non-polarized equilibrium $W = 0$, we find the critical velocities to behave like

$$c^* \approx 1 + \sqrt{F_w \varepsilon} \quad \text{and} \quad c_* \approx 1 - \sqrt{F_w \varepsilon} \tag{20}$$

for small values of the product $F_w \varepsilon$ (Fig. 2).

We note that for the superlinear versions of the alignment term we also find such critical velocities, say $\hat{c}$ and $\check{c}$. However, for given parameters $\alpha$ and $\varepsilon$ these satisfy $\hat{c} > c^*$ and $\check{c} < c_*$. Moreover, the corresponding conditions to (19) are in that case sufficient but not necessary anymore for the existence of monotone waves meaning that these monotone waves might potentially exist also with speeds closer to one.

## 2.2 Traveling Waves for the Full Problem

Returning to the question for traveling wave solutions to the full problem (9), (10) we reduce the system of ordinary differential equations being satisfied by the traveling wave profiles to the first order system

$$U' = Z, \quad \varepsilon Z' = -c Z + V \tag{21}$$
$$W' = V, \quad \varepsilon V' = Z - c V - f(U, W) \tag{22}$$

with $Z$ and $V$ having been introduced as momentum variables.

For $c \neq \pm 1$, system (15), (16) is the reduced system of this full problem in the normally hyperbolic limit $\varepsilon \searrow 0$. According to the singular perturbation theory developed by N. Fenichel in [2] and refined by P. Szmolyan in [4], we can thus assert the existence of heteroclinic orbits connecting the totally aligned equilibria with the non-polarized state or vice versa at least for sufficiently small diffusion coefficients $\varepsilon$ where the meaning of being sufficiently small depends on the wave speed $c$.

The full dynamical system cannot be reduced to an effectively two dimensional problem. We therefore cannot exclude the possibility of further heteroclinic orbits connecting both totally aligned states with one another while passing through the symmetric state $W = 0$ at momenta $Z$ and $V$ being non zero. In the simulations described in the following section we will indeed find this type of wave profiles.

# 3 Traveling Waves Found by Simulations

In this section, we will first discuss which types of traveling wave profiles can be found in the simulations and which wave speeds actually do occur.

## 3.1 Typical Wave Patterns

The easiest wave pattern consists of a single traveling front connecting two equilibria. Given the model parameters and having chosen a wave velocity we can deduce the wave pattern by integrating the system of ordinary differential equations and plugging the result into the simulations as initial condition.

In doing so we observe that the system does not stick to the initial data but selects a wave profile with a distinct velocity which only depends on the model parameters.

In particular, it was not possible to find any parameter setting and initial conditions leading to oscillating wave profiles. As is well known for Fisher-KPP like equations (cf. [5]), these non-monotone fronts seem to be unstable.

In analogy to the patterns obtained by gluing together different wave profiles we observe two types of solutions which typically emerge as long time behavior. Both

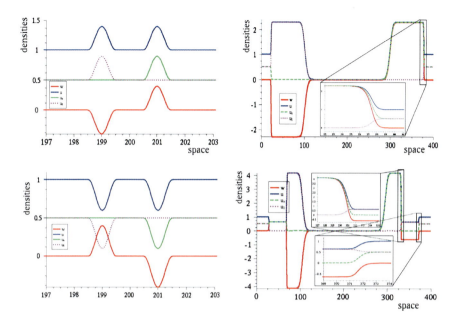

**Fig. 3** Typical initial conditions (*left*) and the traveling wave profiles emerging from them (*right*). *Top* Pattern with a single hump and one wave profile moving to each direction, *bottom* pattern with two traveling wave profiles per direction. In the emerging patterns, in the zoom boxes, the traveling wave profiles of the right moving waves are shown in detail. Note that the pictures of the initial data show only a small region in the center of the domain

of them are characterized by a complete depletion of filaments in the center of the domain and by symmetric equilibria to the far left and to the far right.

Which of these profiles emerges depends on the type of initial conditions. One of these solutions consists of one hump traveling in either direction whereas the second type has two humps per direction—one being totally left aligned, the other one totally right aligned. The two typical initial conditions and the emerging wave patterns are shown in Fig. 3.

Very similar patterns with only one hump moving in only one direction can be observed if the initial data are chosen to by asymmetric.

Let us finally note that the steepness of either traveling wave profile increases with its velocity. Moreover, the profiles are significantly steeper if the alignment term is superlinear as compared to the sublinear version.

## 3.2 Observed Wave Velocities Depending on the Parameters

The first thing to note is that we were not able to find a combination of parameters and initial data leading to a stable wave profile of velocity $|c| < 1$. In fact, the observed traveling waves with the non-polarized equilibrium ($W = 0$, $U_1 > 0$) as one of their asymptotic states have velocities whose absolute value is at least $c^*$. The absolute values of the velocities of the wave profiles connecting two totally aligned states lie between 1 and $c^*$.

Concerning the dependence of the velocities on the parameters $\alpha$ and $\varepsilon$ we observe the following properties which are in good agreement with the predictions we made for the auxiliary problem (13), (10).

1. For any alignment term $f$ and for all observed wave profiles, the wave speed only depends on the product $\alpha\varepsilon$ rather than on both parameters individually.
2. For each type of $f$, we indeed find the absolute velocity $|c|$ to behave like $1 + \sqrt{\alpha\varepsilon}$ at small values of the product $\alpha\varepsilon$ for waves connecting a totally aligned state and the non-polarized state.
3. For given parameters, the superlinear alignment term leads to wave profiles which are steeper and faster than those according to sublinear alignment.

## 4 Conclusion

In order to understand the effect of diffusion and mutual alignment of actin filaments in a minimal model for the cytoskeleton we deduced a basic system of two parabolic equations describing the motion of aligning filaments in one space dimension. For this system, we found different types of traveling wave solutions, depending on the type of alignment and the model parameters $\alpha$ (alignment strength) and $\varepsilon$ (diffusivity of the filaments).

In particular, we found solutions to the system that are composed of different traveling waves and in some cases of additional diffusion profiles. These solutions emerge from minor perturbations of the completely symmetric steady state describing a non-polarized cell at rest which indicates that a small bias in the data can lead to large fronts of filaments as in a cell during directed motion.

Moreover, we found that the velocities of the emerging wave profiles depended on the system parameters and the alignment type in a predictable way.

Recalling the motivation of our model it seems promising to allow for diffusion and alignment of filaments in our cytoskeleton model. We might then expect the formation of fronts of total polarization of the cytoskeleton which can be interpreted as the precursor of lamellipodial structures and actin waves as described in [6]. This is a major challenge for the future as it requires to keep track of the connection between the barbed and pointed end of each filament which should only be possible by incorporating the filament length as an additional variable as it has been done in [1], for instance.

**Acknowledgments** The work of H. Freistühler has been supported by the German Research Foundation (DFG) through its excellence grant to the University of Konstanz. The work of J. Fuhrmann has been supported by the German Federal Ministry of Education and Research through the Bernstein Center for Computational Neuroscience Heidelberg/Mannheim (BmBF, 01GQ1003A) and the DFG through the International Graduate College IGK 710. Part of the research by J.Fuhrmann and A. Stevens was done while they were working at the University of Heidelberg.

# References

1. K. Doubrovinski, K. Kruse, Self-organization in systems of treadmilling filaments. Eur. Phys. J. E. **31**, 95–104 (2010)
2. N. Fenichel, Geometric singular perturbation theory. J. Diff. Eq. **31**, 53–98 (1979)
3. J. Fuhrmann, J. Käs, A. Stevens, Initiation of cytoskeletal asymmetry for cell polarization and movement. J. Theor. Biol. **249**, 278–288 (2007)
4. P. Szmolyan, Transversal heteroclinic and homoclinic orbits in singular perturbation problems. J. Diff. Eq. **92**(2), 252–281 (1991)
5. V. Volpert, S. Petrovskii, Reaction-diffusion waves in biology. Phys. Life Rev. **6**, 267–310 (2009)
6. O.D. Weiner, An actin-based wave generator organizes cell motility. PLoS Biol. **5.9**(e221), 1053–1063 (2007)

# Homing to the Niche: A Mathematical Model Describing the Chemotactic Migration of Hematopoietic Stem Cells

**Maria Neuss-Radu**

**Abstract** It has been shown that hematopoietic stem cells migrate in vitro and in vivo following the gradient of a chemotactic factor produced by stroma cells. In this contribution, a quantitative model for this process is presented. The model consists of chemotaxis equations coupled with an ordinary differential equation on the boundary of the domain and subjected to nonlinear boundary conditions. The existence and uniqueness of a local solution is proved and the model is simulated numerically. It turns out that for adequate parameter ranges, the qualitative behavior of the stem cells observed in the experiment is in good agreement with the numerical results. Our investigations represent a first step in the process of elucidating the mechanism underlying the homing of hematopoietic stem cells quantitatively.

**Keywords** Hematopoietic stem cells · Homing · Chemotaxis equations · Nonlinear boundary conditions

## 1 Introduction

Stem cells are cells with the dual ability to self-renew and to differentiate into multiple cell types. This means that, during the life span of an organism, somatic stem cells give rise to non-self-renewing functionally mature cells, e.g. liver cells, muscle cells, nerve cells, while maintaining a pool of primitive stem cells. Hematopoietic stem cells (HSCs) are the origin of all myeloid/erythroid and lymphoid cell lineages. The natural microenvironment for the HSCs is the stem cell niche in the bone marrow consisting of, among others, stroma cells.

---

M. Neuss-Radu (✉)
Department of Mathematics, University of Erlangen-Nuremberg, Cauerstr. 11,
91058 Erlangen, Germany
e-mail: maria.neuss-radu@math.fau.de

M. Delitala and G. Ajmone Marsan (eds.), *Managing Complexity, Reducing Perplexity*,
Springer Proceedings in Mathematics & Statistics 67, DOI: 10.1007/978-3-319-03759-2_11,
© Springer International Publishing Switzerland 2014

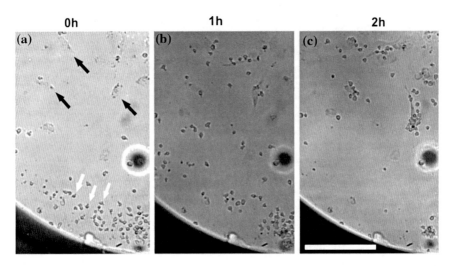

**Fig. 1** HSCs (*white arrows*) were initially seeded on the lower half of the Terasaki well (**a**). They migrated within 2 h toward the stroma cells (*black arrows*) and established cell-cell contact with the stroma cells (**b, c**). From [14]

HSCs are characterized by a rapid migratory activity and their ability to "home" to their niche in the bone marrow. These properties are very important in the therapy of leukemia, which consists mainly of two steps. The first step is a chemotherapy and whole body irradiation to irradicate the patients hematopoietic system. The second step is the transplantation of HSCs obtained from the mobilized peripheral blood of a donor. After transplantation, HSCs find their way of their own accord into the stem cell niche in the bone marrow. Upon homing, the HSCs have to multiply rapidly to regenerate the blood system.

It is a crucial aim of research, to reduce the time necessary for regeneration, a period in which the patient is missing an effective immune system. To achieve this goal, the underlying mechanism of the homing process of the HSCs has to be understood and mathematical models able to quantify this process have to be formulated.

In [1] it was shown that HSCs migrate in vitro and in vivo towards a gradient of the chemotactic factor SDF-1 (stromal cell-derived factor-1) produced by stroma cells. In [14] the experimental assay from Fig. 1 is used to investigate the migration of the stem cells toward the stroma cells.

In this contribution, we describe the migration process observed in [14] quantitatively using a chemotaxis model adapted to our situation. The results presented here are based mainly on the paper [7]. The mathematical model consists of a nonlinear system of two coupled reaction-diffusion equations describing the evolution in time and space of the concentration of stem cells and of the chemoattractant inside the domain, together with an ordinary differential equation (ODE) defined on the part of the boundary coated with stroma cells. This ODE describes the evolution of the stem cells which are attached to the stroma cells. The attachment and detachment of the stem cells at the boundary as well as the production of the chemoattractant by stroma cells are modeled by nonlinear boundary conditions involving the normal fluxes of the stem cells and of the chemoattractant, as well as the concentration of the attached stem cells.

The chemotaxis equations in the classical setting, i.e. with homogeneous Neumann or Dirichlet boundary conditions, have been studied in a large number of papers, a summary of which can be found e.g. in [5]. The solutions may exhibit singularities in finite or infinite time, see e.g. [6, 11]. These singularities model aggregation processes which lead to the formation of $\delta$-functions in the cell concentration. However, in a special case, i.e. for properly chosen sensitivity functions, linear degradation and suitable production of the chemoattractant, in [13] the existence of global weak solutions was proven.

In our paper, similar sensitivity functions as in [13] are used. However, the nonlinear consumption term for the chemoattractant and the nonlinear boundary condition which is new in connection with the chemotaxis equations, require new ideas in the study of the solutions. Here, the existence and uniqueness of a local solution is proven.

The mathematical model formulated in this paper gives a contribution to the quantitative modeling of the homing and engraftment of hematopoietic stem cells. To our knowledge, it is the first model describing the stroma controlled chemotactic migration of HSCs. The simulations in Sect. 4 show that for adequate parameter ranges, the solutions reproduce the qualitative behavior observed in the experiment. Thus, after identifying the relevant parameter of the model using experimental data, we will be able to determine quantitatively the influence which single parameters or combinations of parameters have on the behavior of the HSCs, and thus to provide possibilities to shorten the time needed for homing.

## 2 Mathematical Model

Based on information from the experiment in [14] we set up the following mathematical model describing the chemotactic movement of HSCs, see also [12]. We consider a domain $\Omega \subset \mathbb{R}^2$ of class $C^1$ representing the Terasaki well, see Fig. 2. The boundary of the domain consists of two parts, $\partial\Omega = \Gamma_1 \cup \Gamma_2$, with $\Gamma_1 \cap \Gamma_2 = \emptyset$ and $\Gamma_2$ being a closed subset. The boundary portion $\Gamma_1$ represents the part of the boundary where the stroma cells are cultivated. We denote by $\nu$ the outer unit normal to the bounday $\partial\Omega$. The unknowns of our model are the concentration of the stem cells in the domain $\Omega$ denoted by $s(t, x)$, the concentration of the chemoattractant (SDF-1) denoted by $a(t, x)$, and the concentration of the stem cells bound to stroma cells at the boundary part $\Gamma_1$, denoted by $b(t, x)$.

The evolution of the concentrations $s(t, x), a(t, x)$ is described by the following chemotaxis system

**Fig. 2** The domain $\Omega$ with outer normal vector $\nu$. Stroma cells are cultivated on the boundary portion $\Gamma_1$. No cells or chemoattractant can leave the domain via $\Gamma_2$

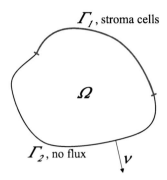

$$\partial_t s = \nabla \cdot (\varepsilon \nabla s - s \nabla \chi(a)), \quad \text{in } (0, T) \times \Omega \tag{1}$$

$$\partial_t a = D_a \Delta a - \gamma a s, \quad \text{in } (0, T) \times \Omega \tag{2}$$

together with the boundary conditions

$$-(\varepsilon \partial_\nu s - s \chi'(a) \partial_\nu a) = \begin{cases} c_1 s - c_2 b, & \text{on } (0, T) \times \Gamma_1 \\ 0, & \text{on } (0, T) \times \Gamma_2 \end{cases} \tag{3}$$

$$D_a \partial_\nu a = \begin{cases} \beta(t, b) c(x), & \text{on } (0, T) \times \Gamma_1 \\ 0, & \text{on } (0, T) \times \Gamma_2. \end{cases} \tag{4}$$

The evolution of the concentration $b(t, x)$ is described by the ODE

$$\partial_t b = c_1 s - c_2 b, \quad \text{on } (0, T) \times \Gamma_1 \tag{5}$$

and $b = 0$ on $(0, T) \times \Gamma_2$. We also impose the initial conditions

$$s(0) = s_0, \ a(0) = a_0 \text{ in } \Omega, \quad \text{and} \quad b(0) = b_0 \text{ on } \Gamma_1. \tag{6}$$

In our model, Eq. (1) describes the random migration of the HSCs, with random motility coeffecient $\varepsilon$, as well as the directional migration in response to the spatial gradient of the chemoattractant. Equation (2) describes the diffusion of the chemoattractant and its consumption due to binding to the receptors expressed on the stem cell membranes. The boundary condition (3) describes the attachment and detachment of stem cells at the part of the boundary coated with stroma cells.

The ODE (5) describes the evolution of the bound stem cells due to the attachment and detachment of stem cells at the boundary $\Gamma_1$.

# 3 Existence and Uniqueness of a Weak Solution

The main result of this section is the existence and uniqueness of a weak solution of the initial boundary value problem (1)–(6). The precise meaning of the concept of weak solution, can be found in [7].

**Theorem 1** *Let the data of our model satisfy the assumptions given in [7], Section 2.1. Then, there exists $T > 0$ and a unique weak solution $(s, a, b)$ of the system (1)–(6). This solution is positive and has the additional regularity properties $a \in L^2(0, T; H^2(\Omega)) \cap C([0, T]; H^1(\Omega)) \cap L^\infty(0, T; L^\infty(\Omega))$ and $b \in C([0, T]; L^2(\partial\Omega))$.*

The proof of Theorem 1 consists of several steps. First, we cut off the concentration $s$ of stem cells in the nonlinear terms. Using a fixed point argument, we prove the existence of a solution for the resulting system and afterwards, we show uniqueness and positivity of this solution. Finally, we prove that the concentration of stem cells in the cut-off system is bounded, so that this solution is the solution of our original system as well. For a detailed proof, see [7], Theorem 1.

# 4 Numerical Results

In this section, we present numerical simulations for our model (1)–(6). The tool kit Gascoigne, see www.gascoigne.de, is used. The simulation is realized on the rectangle $(0, 1, 5) \times (0, 1)$ with a grid with $129 \times 65$ nodes.

The stroma cells are concentrated on the right boundary ($x_1 = 1, 5$) where they are mainly distributed in three clusters. A precise description of the distribution of stroma cells and of the function $\beta(t, b)$ in the production rate of the chemoattractant is given in [7]. For the simulation, we consider a linear sensitivity function $\chi(a) = 10a$. We choose as initial conditions $a_0 = 0$, $b_0 = 0$ and

$$s_0(x_1, x_2) = \begin{cases} (1 + \cos(5\pi(x_1 - 0, 4))) \sin(\pi x_2) & \text{for } 0, 2 \le x_1 \le 0, 6 \\ 0 & \text{otherwise,} \end{cases}$$

See also Fig. 3. The Figs. 4, 5, and 6 describe the time evolution of the solution components $s$ and $b$.

# 5 Discussion

In this contribution, we give a quantitative model for the movement of HSCs in chemotactic gradients, based on the experimental results from [14], in a close collaboration with the stem cell research group of Prof. Ho (Medical Clinic, University

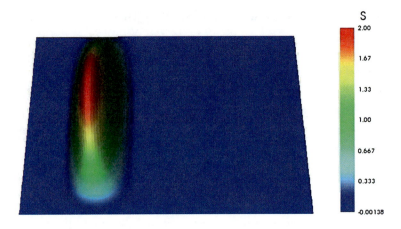

**Fig. 3** Initial concentration of the stem cells $s_0$

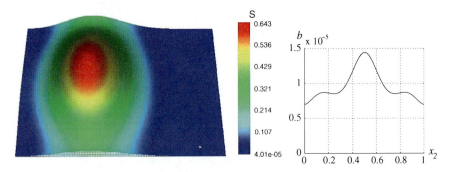

**Fig. 4** The free and bound stem cells $s$ and $b$ at time $t = 10$

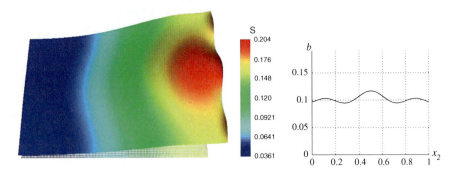

**Fig. 5** The free and bound stem cells $s$ and $b$ at time $t = 45$

of Heidelberg). This is a first step in the process of elucidating the mechanism underlying the homing of HSCs quantitatively. Methods enabling us to control the homing process, the motility and motion of stem cells are highly valuable for medical reasons. In the therapy of some forms of leukemia the ability of HSCs to home into

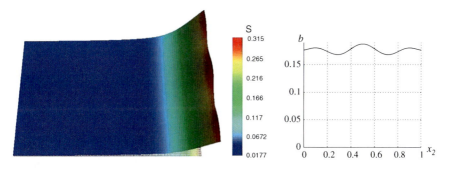

**Fig. 6** The free and bound stem cells $s$ and $b$ at time $t = 100$

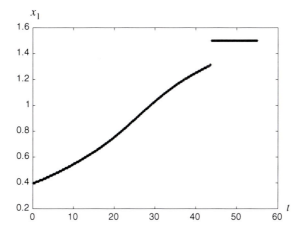

**Fig. 7** Variation in time of the position of the maximum of the HSCs' concentration. The $x_1$-coordinate is plotted. Due to the symmetry of the problem the $x_2$-coordinate is constant, equal to 0.5. The jump in the position of the maximum, seen after 438 time steps, is due to the accumulation of stem cells at the boundary $x_1 = 1, 5$ and the formation of the absolute maximum at the boundary

their niche in the bone marrow is of utmost importance for the regeneration of the blood system. It is a crucial aim to reduce the time necessary for regeneration, and thus the risk for the patient and the costs for the health system.

So far, the mechanisms and specific molecules involved in the homing process are still not fully understood. However, there is evidence that human $CD34^+/38^-$ stem cells are attracted by stromal cell-derived factor-1 (SDF1), a chemoattractant produced by bone marrow stromal cells, see [1, 14].

The mathematical modelling may help to order the achieved experimental results and to pose structured questions to the experimentalists. It played already a substantial role in designing experiments by suggesting and selecting possible factors and mechanisms, which are important for a quantitative description. E.g. for our model, the random motility coefficient and the chemotactic sensitivity are two important parameters. Whereas the random motility coefficient can be measured directly, e.g. by measuring migration in a uniform concentration of the chemoattractant, the chemotactic sensitivity is difficult to measure directly. With the help of our model, we can determine numerically if the chemotactic sensitivity and the migration velocity of the HSCs are correlated. In case of a strong correlation, an experiment for measuring the migration velocity has to be designed. For the example considered in Sect. 4, the migration velocity can be measured by the slope of the position of the maximum of HSCs' concentration with respect to time, see Fig. 7.

Experimental research, taking into account results of our modelling and simulations, are just going to provide the data needed for the calibration. After calibration the model can be used for computer experiments. Furthermore, we remark that the model derived here is representing a larger class of systems modelling spread in space and time controlled by processes on the boundary.

# References

1. A. Aiuti, I.J. Webb, C. Bleul, T. Springer, J.C. Gutierrez-Ramos, The chemokine sdf-1 is a chemoattractant for human $cd34^+$ hematopoietic progenitor cells and provides a new mechanism to explain the mobilization of $cd34^+$ progenitors to peripheral blood. J. Exp. Med. **185**, 111–120 (1997)
2. L. Evans, *Partial Differential Equations* (AMS, Providence, 1999)
3. H. Gajewski, K. Gröger, K. Zacharias, *Nichtlineare Operatorgleichungen und Operatordifferentialgleichungen* (Akademie-Verlag, Berlin, 1974)
4. P. Grisvard, *Elliptic Problems in Nonsmooth Domains. Monographs and Studies in Mathematics*, vol. 24 (Pitman, Boston, 1985)
5. D. Horstmann, From 1970 until present: the keller-segel model in chemotaxis and its consequences. Jahresbericht der DMV **105**, 103–165 (2004)
6. W. Jäger, S. Luckhaus, On explosions of solutions to a system of partial differential equations. Trans. AMS **329**, 819–824 (1992)
7. A. Kettemann, M. Neuss-Radu, Derivation and analysis of a system modeling the chemotactic movement of hematipoietic stem cells. J. Math. Biol. **56**, 579–610 (2008)
8. O. Ladyženskaya, V. Solonnikov, N. Ural'ceva, Linear and quasilinear equations of parabolic type. AMS Trans. Math. Monogr. **23**, 179 (1968)
9. J. Lions, *Quelque méthodes de résolution des problèmes aux limites non linéaires* (Dunod, Paris, 1969)
10. J. Lions, E. Magenes, *Non-Homogeneous Boundary Value Problems and Applications*, vol. 1 (Springer, Berlin, 1972)
11. T. Nagai, Blow-up of radially symmetric solutions to a chemotaxis system. Adv. Math. Sci. Appl. **5**, 581–601 (1995)
12. M. Neuss-Radu, A. Kettemann, A mathematical model for stroma controlled chemotaxis of hematopoietic stem cells. Oberwolfach Reports **24**, 59–62 (2006)
13. K. Post, *A system of non-linear partial differential equations modeling chemotaxis with sensitivity functions.* Humboldt Universität zu Berlin, Mathematisch-Naturwissenschaftliche Fakultät II, electronic (1999)
14. W. Wagner, R. Saffrich, U. Wrikner, V. Eckstein, J. Blake, A. Ansorge, C. Schwager, F. Wein, K. Miesala, W. Ansorge, A.D. Ho, Hematopoietic progenitor cells and cellular microenvironment: behavioral and molecular changes upon interaction. Stem Cells **23**, 1180–1191 (2005)

# DDE Models of the Glucose-Insulin System: A Useful Tool for the Artificial Pancreas

**Jude D. Kong, Sreedhar S. Kumar and Pasquale Palumbo**

**Abstract**  Delay differential equations are widely adopted in life sciences: including delays explicitly in mathematical models allows to simulate the systems under investigation more accurately, without the use of auxiliary fictitious compartments. This work deals with Delay Differential Equation (DDE) models exploited in the specific framework of the glucose-insulin regulatory system, and a brief review of the DDE models available in the literature is presented. Furthermore, recent results on the closed loop control of plasma glycemia, based on DDE models of the individual glucose-insulin system are summarized. Indeed, DDE models revealed to be particularly suited to simulate the pancreatic insulin delivery rate, thereby allowing to treat in a unified fashion both Type 1, where no endogenous insulin release is available, and Type 2 diabetic patients, where the exogenous insulin administration adds up to the endogenous insulin production.

**Keywords**  Glucose-insulin system · Delay differential equation models · Artificial pancreas

---

J. D. Kong (✉)
Department of Mathematical and Statistical Sciences, University of Alberta, 562 Central
Academic Building, Edmonton, AB  T6G 2G1, Canada
e-mail: jdkong@ualberta.ca

S. S. Kumar
Department of Microsystems Engineering (IMTEK), Faculty of Engineering, Albert-Ludwigs
University of Freiburg, Freiburg, Germany
e-mail: sreedhar.kumar@imtek.de

P. Palumbo
BioMatLab, CNR-IASI, UCSC, Largo A. Gemelli 8, 00168 Rome, Italy
e-mail: pasquale.palumbo@iasi.cnr.it

M. Delitala and G. Ajmone Marsan (eds.), *Managing Complexity, Reducing Perplexity*,
Springer Proceedings in Mathematics & Statistics 67, DOI: 10.1007/978-3-319-03759-2_12,
© Springer International Publishing Switzerland 2014

# 1 Introduction

Diabetes Mellitus comprises a group of metabolic diseases characterized by hyperglycemia. The chronic hyperglycemia of diabetes is associated with long-term damage, dysfunction, and failure of different organs, especially eyes, kidneys, nerves, heart, and blood vessels. Patients with diabetes have an increased incidence of atherosclerotic cardiovascular, peripheral vascular, and cerebrovascular diseases. Diabetes a very high incidence: the number of diabetic patients is expected to double by the year 2030, compared to 2000 data [1]. Hence, diabetes management has a heavy impact on many national public health budgets.

In a healthy person, the blood glucose is maintained between 3.9 and 6.9 mmol/L by means of a complex control system which ensures a balance between glucose entering the bloodstream after liver gluconeogenesis and intestinal absorption following meals, and glucose uptake from the peripheral tissues. This balance is regulated mainly by the insulin, a hormone produced by the $\beta$-cells of the pancreas when properly stimulated by the level of plasma glycemia: indeed, insulin enhances the glucose uptake in the muscles and the adipose tissues as well as it promotes the stocking of circulating glucose in excess to the liver.

A pathological increase in blood glucose concentration (hyperglycemia) results from defects in insulin secretion, insulin action, or both. In case of an absolute deficiency of insulin secretion, caused by an autoimmune destruction of the pancreatic $\beta$ cells, Type 1 diabetes occurs: these patients require exogenous insulin administration for survival. On the other hand, in case of hyperglycemia caused by a combination of resistance to insulin action and inadequate compensatory insulin secretory response, Type 2 diabetes occurs: these patients have therefore insulin resistance and usually also a relative (rather than absolute) insulin deficiency, in the face of increased levels of circulating glucose.

The basic therapeutic procedure for diabetes is the exogenous administration of insulin. This compensation could be accomplished by means of a variety of schemes, depending on the *a priori* knowledge of the patient's glucose-insulin homeostasis and on the technology available for actuating the designed control law. In most widespread cases, glucose control strategies are mainly actuated by subcutaneous administration of insulin, with the dose adjusted by the patients themselves, on the basis of capillary plasma glucose concentration measurements. On the other hand, a real-time closed-loop control scheme would require an algorithm that provides the proper dose of the hormone independently of any action on the patient, and is robust with respect to the many sources of perturbation of the glucose-insulin system like meal ingestion, physical exercise or trivially, malfunctioning of the artificial pancreas devices. To this aim, the use of a mathematical model of the patient's glucose-insulin system would allow to exploit individual optimal strategies to synthesize the exogenous insulin administration. Clearly, the more accurate the model, the more efficient will the control law be.

The modeling of the glucose-insulin system is an appealing and challenging topic in biomathematics and many different models have been presented in the last decades (see e.g. [2, 3] and references therein). Section 2 provides a brief review on Delay Differential Equation (DDE) models of the glucose-insulin system, and aims to motivate the reason because so many DDE models appeared in the literature along the past decade; Sect. 3 presents recent results on DDE-model-based control laws for the artificial pancreas, focusing on the state of the art, main results obtained and future developments.

## 2 DDE Models of the Glucose-Insulin System

Most of the available glucose-insulin models are strongly related to the experimental framework they want to replicate and can be roughly split into two main branches: the ones concerning short period experiments like, e.g., the IntraVenous/Oral Glucose Tolerance Test (IVGTT/OGTT), that last no more than 5/6 h, and the others related to long period experiments, mainly concerning the glucose/insulin ultradian oscillations, that usually last 24 h.

### 2.1 Short Period DDE Models

As far as short period experiments are concerned, models have been proposed mainly with the purpose of estimating the individual insulin sensitivity of tissues in order to predict a possible diabetes progression. In this framework, the mostly used model in physiological research of the glucose metabolism is the Minimal Model [4, 5], proposed for the interpretation of the IVGTT. It consists of three coupled ordinary differential equations, one for the insulin and two for the glucose dynamics, modeling the apparent delay of insulin action on the insulin-dependent glucose uptake by means of an auxiliary remote compartment. The Minimal Model played a crucial role in modeling the glucose-insulin system, mainly because it provided the insulin sensitivity as a combination of the model parameter, thus coming out as a by-product of the model identification procedure. However, some criticisms have been raised in the last decade, mainly related to the mathematical coherence of the model (the coupled equations do not ensure bounded solutions, nor a steady-state equilibrium) and to the lack of apparent validity besides the IVGTT experimental framework.

First DDE models of the glucose-insulin system have been actually proposed to overcome these drawbacks. In [6] the Authors deleted the remote compartment in the glucose dynamics and introduced a distributed delay for the glucose-dependent Insulin Delivery Rate (IDR). Besides being mathematically coherent and more versatile to different sets of experiments apart from the IVGTT, such a DDE model has also been validated on real data and, moreover, it provides the insulin sensitivity by the estimate of a single parameter. Thereafter, there has been a widespread development of DDE models, which revealed to be particularly suitable to replicate the IDR. For instance families of DDE models have been proposed, where general delays are introduced both in the insulin action on tissue glucose uptake and in the glucose action on pancreatic insulin secretion, [7, 8].

Despite the development of different DDE models of the glucose-insulin system, they have not been adopted in a model-based framework for the artificial pancreas, since [9]. This is because most of the efforts in this research area have been mainly devoted to Type 1 diabetic patients, in whom the absence of a pancreatic IDR motivates urgent research efforts in closed-loop exogenous insulin infusion therapies, and weaken the necessity of preferring DDE models instead of ODE ones. On the other hand, the ability of time-delay systems to better model the endogenous IDR makes it so that DDE-model-based approaches could reveal to be very effective for treating the much more prevalent category of Type 2 diabetic patients.

Below are reported the equations of a DDE model recently exploited for theoretical research in artificial pancreas [10]

$$\frac{dG(t)}{dt} = -K_{xgi}G(t)I(t) + \frac{T_{gh}}{V_G},$$

$$\frac{dI(t)}{dt} = -K_{xi}I(t) + \frac{T_{iGmax}}{V_I}f(G(t - \tau_g)), \qquad f(G) = \frac{(\frac{G}{G^*})^\gamma}{1 + (\frac{G}{G^*})^\gamma}. \qquad (1)$$

where $G(t)$, [mM] and $I(t)$, [pM], denote plasma glycemia and insulinemia. $K_{xgi}$, [min$^{-1}$ pM$^{-1}$], is the rate of glucose uptake by insulin-dependent tissues per pM of plasma insulin concentration; $T_{gh}$, [min$^{-1}$ (mmol/kgBW)], is the net balance between hepatic glucose output and insulin-independent zero-order glucose tissue uptake; $V_G$ and $V_I$, [L/kgBW], are the apparent glucose and insulin distribution volume; $K_{xi}$, [min$^{-1}$], is the apparent first-order disappearance rate constant for insulin; $T_{iGmax}$, [min$^{-1}$(pmol/kgBW)], is the maximal rate of second-phase insulin release; $\tau_g$, [min], is the apparent delay with which the pancreas varies secondary insulin release in response to varying plasma glucose concentrations; $\gamma$ is the progressivity with which the pancreas reacts to circulating glucose concentrations and $G^*$, [mM], is the glycemia at which the insulin release is half its maximal rate.

Mathematical coherence has been proven in [8], where the model has been shown to provide positive and bounded solutions, and is endowed with a unique asymptotically stable equilibrium point (for basal glycemia and insulinemia). Sufficient conditions are also given for global stability, that has been investigated in further papers [11, 12].

## 2.2 Long Period DDE Models

Long-term models of the glucose-insulin system are mainly motivated to reproduce the phenomenon of sustained, apparently regular, long period oscillations of glycemia and insulinemia, known as *ultradian* oscillations. A pioneering work in such a framework has been the paper of Sturis et al. in 1991 [13], a sixth order nonlinear ODE model according to which the Authors proposed a plausible mechanism for the genesis of the oscillations, suggesting they could originate from the glucose-insulin reciprocal interaction without postulating an intra-pancreatic pacemaker for their existence. In fact, the model presents two delays, both realized by means of additional fictitious compartments: one delay is associated to the suppression of glucose production by insulin (two-compartment model for the insulin kinetics), while the other is related to the effect of insulin on glucose production (four compartment model for the glucose kinetics). The model of Sturis et. al. has been the starting point for many further DDE models, aiming to replicate the occurrence of long period oscillations as coming from a Hopf bifurcation point (see, e.g. [14–18]). It has to be stressed that though the model of Sturis et. al. and its DDE versions have been used, especially in recent years, to study the effect of pulsatile insulin profiles in (pre)-diabetic patients [19–23], to the best of the authors' knowledge, they have not yet been adopted to synthesize a model-based control law for insulin therapy.

## 3 DDE Model Based Control

First results on DDE-model-based control of the glucose-insulin system can be found in [9, 24], where the DDE model described in (1) was considered for a possible intravenous (iv) administration of the insulin therapy. To this aim the insulin equation in (1) is endowed with an additive control input $u(t)$. Compared to the usual subcutaneous insulin injection, the use of iv insulin administration, delivered by automatic, variable speed pumps provides a wider range of possible strategies and ensures a rapid delivery with negligible delays. As a matter of fact, control algorithms based on iv insulin administration are directly applicable so far only to problems of glycemia stabilization in critically ill subjects, such as in surgical intensive care units after major procedures.

In [9, 24] the input-output linearization with delay cancelation is achieved, by means of suitable inner and outer feedback control laws, with guaranteed internal stability. In particular, a reliable, causal state feedback which allows to reduce a high basal plasma glucose concentration to a lower level, according to a smooth reference glucose trajectory $G_{\text{ref}}(t)$, is designed with:

$$u(t) = \frac{S(G(t), I(t), G(t - \tau_g)) - v(t)}{K_{xgi} G(t)} \tag{2}$$

where

$$S(G(t), I(t), G(t - \tau_g)) = -K_{xgi} I(t) \left( -K_{xgi} I(t) G(t) + \frac{T_{gh}}{V_G} \right) \tag{3}$$
$$- K_{xgi} G(t) \left( -K_{xgi} I(t) + \frac{T_{iGmax}}{V_I} f(G(t - \tau_g)) \right)$$

and $v(t) = \ddot{G}_{ref}(t) + Re(t)$, with $R \in \mathbb{R}^{1 \times 2}$ a matrix such that

$$H = \begin{bmatrix} 0 & 1 \\ 0 & 0 \end{bmatrix} + \begin{bmatrix} 0 \\ 1 \end{bmatrix} R \tag{4}$$

has prescribed eigenvalues with negative real part and $e(t) = Z(t) - Z_{ref}(t)$, with

$$Z(t) = \begin{bmatrix} z_1(t) \\ z_2(t) \end{bmatrix} = \begin{bmatrix} G(t) \\ -K_{xgi}G(t)I(t) + \frac{T_{gh}}{V_G} \end{bmatrix}, \qquad Z_{ref}(t) = \begin{bmatrix} G_{ref}(t) \\ \dot{G}_{ref}(t) \end{bmatrix} \tag{5}$$

The glucose reference signal to be tracked, $G_{ref}(t)$, is supposed to be bounded, twice continuously differentiable, with bounded first and second derivatives. Such a closed-loop control law ensures input-to-state stability of the closed loop error system with respect to disturbances occurring in the insulin dynamics, such as insulin actuator malfunctions.

The main drawback concerns the necessity to exploit both glucose and insulin measurement at the present and at a delayed time: insulin measurements are slower and more cumbersome to obtain, more expensive, and also less accurate than glucose measurements. A need exists, therefore, to design a control law avoiding real-time insulin measurements. To this aim, in order to close the loop by means of only glucose measurements, a state observer for the DDE system (1) has been proposed in [25, 26]. By suitably exploiting the state observer theory for nonlinear time delay systems (see [27]), the observer equations for the estimates of glycemia and insulinemia , $\hat{G}(t)$, and $\hat{I}(t)$ respectively, are given by

$$\begin{bmatrix} \frac{d\hat{G}(t)}{dt} \\ \frac{d\hat{I}(t)}{dt} \end{bmatrix} = \begin{bmatrix} -K_{xgi}\hat{G}(t)\hat{I}(t) + \frac{T_{gh}}{V_G} \\ -K_{xi}\hat{I}(t) + \frac{T_{iGmax}}{V_I} f\big(\hat{G}(t - \tau_g)\big) + u(t) \end{bmatrix} + Q^{-1}(\hat{G}(t), \hat{I}(t))W(G(t) - \hat{G}(t)), \tag{6}$$

where $Q^{-1} \in \mathbb{R}^{2 \times 2}$ is the inverse matrix of the Jacobian of the observability map (see [28]), here given, for $\begin{bmatrix} x_1 \\ x_2 \end{bmatrix} \in \mathbb{R}^2$, by $\begin{bmatrix} x_1 \\ -K_{xgi}x_1 x_2 + \frac{T_{gh}}{V_G} \end{bmatrix}$. The gain matrix $W \in \mathbb{R}^{2 \times 1}$ is chosen in order to assign suitable eigenvalues to matrix $\hat{H}$, defined by means of the Brunowski pair $(A_b, C_b)$ as

$$\hat{H} = A_b - W C_b, \quad \text{where} \quad A_b = \begin{bmatrix} 0 & 1 \\ 0 & 0 \end{bmatrix}, \quad C_b = \begin{bmatrix} 1 & 0 \end{bmatrix}. \tag{7}$$

In order to close the loop from the observed state, the control law (2)–(5) suitably exploits the estimates $\hat{G}$ and $\hat{I}$ as follows

$$u(t) = \frac{S(\hat{G}(t), \hat{I}(t), \hat{G}(t - \tau_g)) - v(t)}{K_{xgi}\hat{G}(t)}, \quad t \geq 0 \tag{8}$$

DDE Models of the Glucose-Insulin System: A Useful Tool for the Artificial Pancreas    115

with $v(t) = \ddot{G}_{ref}(t) + R\hat{e}(t)$, $\hat{e}(t) = \hat{Z}(t) - Z_{ref}(t)$, and

$$\hat{Z}(t) = \begin{bmatrix} \hat{z}_1(t) \\ \hat{z}_2(t) \end{bmatrix} = \begin{bmatrix} \hat{G}(t) \\ -K_{xgi}\hat{G}(t)\hat{I}(t) + \frac{T_g h}{V_G} \end{bmatrix} \tag{9}$$

Such a control law has been proven to ensure the local convergence of the tracking error to zero. Simulations that validated the theoretical results were also performed in a virtual environment, showing that the results are robust with respect to a wide range of parameter uncertainties or device malfunction. Additionally, the control law has been further evaluated by closing the loop on a *virtual patient*, whose model equations are different from the ones used to synthesize the control law [29]. That means: a minimal model of the glucose-insulin system to design the insulin therapy, and a different, more detailed, comprehensive model to test in silico the control scheme. Such a chosen *maximal* model for the virtual patient, [30], has been recently accepted by the Food and Drug Administration (FDA) as a substitute to animal trials for the preclinical testing of control strategies in artificial pancreas.

Further developments on such a research line involve subcutaneous insulin administration, instead of intravenous infusions, that are usually provided under the direct supervision of a physician. To this aim, in [31–33], the model Eq. (1) are coupled to simple linear model of the insulin absorption from the subcutaneous depot, already exploited with the aim of glucose control in [34]:

$$\frac{dG}{dt} = -K_{xgi}G(t)I(t) + \frac{T_g h}{V_G},$$

$$\frac{dI}{dt} = -K_{xi}I(t) + \frac{T_{iGmax}}{V_I} f\left(G(t - \tau_g)\right) + \frac{S_2(t)}{V_I t_{max,I}},$$

$$\frac{dS_2}{dt} = \frac{1}{t_{max,I}} S_1(t) - \frac{1}{t_{max,I}} S_2(t),$$

$$\frac{dS_1}{dt} = -\frac{1}{t_{max,I}} S_1(t) + u(t), \tag{10}$$

with $t_{max,I}$, [min], the time-to-maximum insulin absorption. The same ideas based on the input/output feedback linearization are applied in this framework with, however, much more complicated formulas to synthesize the control law: preliminary results can be found in [32] where the control law is synthesized by assuming a complete knowledge of the state of the system (i.e. glucose and insulin real-time measurements), and local convergence to zero of the tracking error $G(t) - G_{ref}(t)$ is proven. In [33] the same convergence results are obtained by means of a state observer for the intravenous and subcutaneous insulin values, and the convergence to zero of the tracking error is proven in [31].

# References

1. S. Wild, G. Roglic, A. Green, R. Sicree, H. King, Global prevalence of diabetes: estimates for the year 2000 and projections for 2030. Diabetes Care **27**, 1047–1053 (2004)
2. C. Cobelli, C. Dalla Man, G. Sparacino, L. Magni, G. De Nicolao, B.P. Kovatchev, Diabetes: models, signals, and control. IEEE Rev. Biomed. Eng. **2**, 54–96 (2009)
3. A. Makroglou, J. Li, Y. Kuang, Mathematical models and software tools for the glucose-insulin regulatory system and diabetes: an overview. Appl. Numer. Math. **56**, 559–573 (2006)
4. R.N. Bergman, Y.Z. Ider, C.R. Bowden, C. Cobelli, Quantitative estimation of insulin sensitivity. Am. J. Physiol. **236**, E667–E677 (1979)
5. G. Toffolo, R.N. Bergman, D.T. Finegood, C.R. Bowden, C. Cobelli, Quantitative estimation of beta cells sensitivity to glucose in the intact organism: a minimal model of insulin kinetics in dog. Diabetes **29**, 979–990 (1980)
6. A. De Gaetano, O. Arino, Mathematical modelling of the intravenous glucose tolerance test. J. Math. Biol. **40**, 136–168 (2000)
7. J. Li, Y. Kuang, Analysis of IVGTT glucose-insulin interaction models with time delay. Discrete Continuous Dyn. Syst. Ser. B **1**(1), 103–124 (2001)
8. P. Palumbo, S. Panunzi, A. De Gaetano, Qualitative behavior of a family of delay-differential models for the glucose-insulin system. Discrete Continuous Dyn. Syst. Ser. B **7**(2), 399–424 (2007)
9. P. Palumbo, P. Pepe, S. Panunzi, A. De Gaetano, Robust closed-loop control of plasma glycemia: a discrete-delay model approach, in *Proceedings of 47th IEEE Conference on Decision and Control*, Cancun, Mexico, pp. 3330–3335 (2008)
10. S. Panunzi, P. Palumbo, A. De Gaetano, A discrete single delay model for the intra-venous glucose tolerance test. Theor. Biol. Med. Model. **4**, 35 (2007)
11. D.V. Giang, Y. Lenbury, A. De Gaetano, P. Palumbo, Delay model of glucose-insulin systems: global stability and oscillated solutions conditional on delays. J. Math. Anal. Appl. **343**, 996–1006 (2008)
12. J. Li, M. Wang, A. De Gaetano, P. Palumbo, S. Panunzi, The range of time delay and the global stability of the equilibrium for an IVGTT model. Math. Biosci. **235**, 128–137 (2012)
13. J. Sturis, K.S. Polonsky, E. Mosekilde, E. Van Cauter, Computer model for mechanisms underlying ultradian oscillations of insulin and glucose. Am. J. Physiol. **260**, 439–445 (1991)
14. D.L. Bennett, S.A. Gourley, Asymptotic properties of a delay differential equation model for the interaction of glucose with plasma and interstitial insulin. Appl. Math. Comput. **151**, 189–207 (2004)
15. A. Drozdov, H. Khanina, A model for ultradian oscillations of insulin and glucose. Math. Comput. Model. **22**, 23–38 (1995)
16. K. Engelborghs, V. Lemaire, J. Bélair, D. Roose, Numerical bifurcation analysis of delay differential equations arising from physiological modeling. J. Math. Biol. **42**, 361–385 (2001)
17. J. Li, Y. Kuang, C.C. Mason, Modeling the glucose-insulin regulatory system and ultradian insulin secretory oscillations with two explicit time delays. J. Theor. Biol. **242**, 722–735 (2006)
18. J. Li, Y. Kuang, Analysis of a model of the glucose-insulin regulatory system with two delays. SIAM J. Appl. Math. **67**(3), 757–776 (2007)
19. C.-L. Chen, H.-W. Tsai, Modeling of physiological glucose-insulin system on normal and diabetic subjects. Comput. Methods Programs Biomed. **97**, 130–140 (2010)
20. C.-L. Chen, H.-W. Tsai, S.-S. Wong, Modeling of physiological glucose-insulin dynamic system on diabetics. J. Theor. Biol. **265**, 314–322 (2010)
21. H. Wang, J. Li, Y. Kuang, Mathematical modelling and qualitative analysis of insulin therapies. Math. Biosci. **210**, 17–33 (2007)
22. H. Wang, J. Li, Y. Kuang, Enhanced modelling of the glucose-insulin system and its application in insulin therapies. J. Biol. Dyn. **3**(1), 22–38 (2009)
23. Z. Wu, C.-K. Chui, G.-S. Hong, S. Chang, Physiological analysis on oscillatory behavior of glucose-insulin regulation by model with delays. J. Theor. Biol. **280**, 1–9 (2011)

24. P. Palumbo, P. Pepe, S. Panunzi, A. De Gaetano, Robust closed-loop control of plasma glycemia: a discrete-delay model approach. Discrete Continuous Dyn. Syst. Seri. B (Special Issue on Mathematical Biology and Medicine) **12**(2), 455–468 (2009)
25. P. Palumbo, P. Pepe, S. Panunzi, A. De Gaetano, Observer-based closed-loop control of plasma glycemia, in *Proceedings of 48th IEEE Conference on Decision and Control*, Shanghai, China, pp. 3507–3512 (2009)
26. P. Palumbo, P. Pepe, S. Panunzi, A. De Gaetano, Time-delay model-based control of the glucose-insulin system, by means of a state observer. Eur. J. Control **18**(6), 591–606 (2012)
27. A. Germani, P. Pepe, A state observer for a class of nonlinear systems with multiple discrete and distributed time delays. Eur. J. Control **11**(3), 196–205 (2005)
28. A. Germani, C. Manes, P. Pepe, An asymptotic state observer for a class of nonlinear delay systems. Kybernetika **37**(4), 459–478 (2001)
29. P. Palumbo, G. Pizzichelli, S. Panunzi, P. Pepe, A. De Gaetano, Tests on a virtual patient for an observer-based, closed-loop control of plasma glycemia, in *50th IEEE Conference on Decision and Control & 11th European Control Conference (CDC-ECC 2011)*, pp. 6936–6941 (2011)
30. C. Dalla Man, R.A. Rizza, C. Cobelli, Meal simulation model of the glucose-insulin system. IEEE Trans. Biomed. Eng. **54**(10), 1740–1749 (2011)
31. P. Palumbo, P. Pepe, J.D. Kong, S.S. Kumar, S. Panunzi, A. De Gaetano, Regulation of the human plasma glycemia by means of glucose measurements and subcutaneous insulin administration. American Control Conference (2013) (Submitted)
32. P. Palumbo, P. Pepe, S. Panunzi, A. De Gaetano, Glucose control by subcutaneous insulin administration: a DDE modelling approach, in *Proceedings of 18th IFAC World Congress*, Milan, pp. 1471–1476 (2011)
33. P. Palumbo, P. Pepe, S. Panunzi, A. De Gaetano, Observer-based glucose control via subcutaneous insulin administration, in *Proceedings of the 8th IFAC Symposium on Biological and Medical Systems*, Budapest (2012)
34. R. Hovorka, V. Canonico, L.J. Chassin, U. Haueter, M. Massi-Benedetti, M.O. Federici, T.R. Pieber, H.C. Shaller, L. Schaupp, T. Vering, M.E. Wilinska, Nonlinear model predictive control of glucose concentration in subjects with type I diabetes. Physiol. Meas. **25**, 905–920 (2004)

# Physics and Complexity: An Introduction

**David Sherrington**

**Abstract** Complex macroscopic behaviour can arise in many-body systems with only very simple elements as a consequence of the combination of competition and inhomogeneity. This paper attempts to illustrate how statistical physics has driven this recognition, has contributed new insights and methodologies of wide application, influencing many fields of science, and has been stimulated in return.

**Keywords** Complex systems · Spin glasses · Hard optimization · Neural networks · Econophysics · Conceptual transfers through mathematics

## 1 Introduction

Many body systems of even very simple microscopic constituents with very simple interaction rules can show novel emergence in their macroscopic behaviour. When the interactions (and any constraints) are also mutually incompatible (frustrated) and there is macroscopically relevant quenched disorder, then the emergent macroscopic behaviour can be complex (in ways to be discussed) and not simply anticipatable. Recent years have seen major advances in understanding such behaviour, in recognizing conceptual ubiquities across many apparently different systems and in forging, transferring and applying new methodologies. Statistical physics has played a major part in driving and developing the subject and in providing new methods to study and quantify it. This paper is intended to provide a brief broadbrush introduction.

A key part of these developments has been the combination of minimalist modelling, development of new concepts and techniques, and fruitful transfers of

---

D. Sherrington (✉)
Rudolf Peierls Centre for Theoretical Physics, 1 Keble Rd., Oxford OX1 3NP, UK
e-mail: D.Sherrington1@physics.ox.ac.uk

D. Sherrington
Santa Fe Institute, 1399 Hyde Park Rd., Santa Fe, NM 87501, USA

M. Delitala and G. Ajmone Marsan (eds.), *Managing Complexity, Reducing Perplexity*,
Springer Proceedings in Mathematics & Statistics 67, DOI: 10.1007/978-3-319-03759-2_13,
© Springer International Publishing Switzerland 2014

120                                                                                    D. Sherrington

the knowledge between different systems. Here we shall concentrate on a simple paradigmic model, demonstrate its ubiquitousness among several often very different systems, problems and contexts, and introduce some of the useful concepts that have arisen.

## 2 The Dean's Problem and Spin Glasses

The genesis for the explosion of interest and activity in complexity within the physics community was in an attempt to understand a group of magnetic alloys known as spin glasses[1] [12]. But here we shall start with a problem that requires no physics to appreciate, the Dean's problem [8].

A College Dean is faced with the task of distributing $N$ students between two dormitories as amicably as possible but given that some pairs of students prefer to be in the same dorm while other pairs want to be separated. If any odd number of students have an odd number of antagonistic pairwise preferences then their preferences cannot all be satisfied simultaneously. This is an example of frustration. The Dean's Problem can be modelled as a mathematical optimization problem by defining a cost function $H$ that is to be minimiized:

$$H = -\sum_{(ij)} J_{ij}\sigma_i\sigma_j; \sigma = \pm 1 \tag{1}$$

where the $i, j$ label students, $\sigma = \pm 1$ indicates dorm A/B and the $\{J_{ij} = J_{ji}\}$ denote the sign and magnitude of the inter-student pair preferences (+ = prefer). We shall further concentrate on the situation where the $\{J_{ij}\}$ are chosen randomly and independently from a single (intensive) distribution $P(J)$ of zero mean[2], the random Dean's Problem. The number of combinations of possible choices grows exponentially in $N$ (as $2^N$). There is also, in general, no simple local iterative mode of solution. Hence, in general, when $N$ becomes large the Dean's problem becomes very hard, in the language of computer science NP-complete [6].

In fact, the cost function of the random Dean's Problem was already introduced in 1975 as a potentially soluble model for a spin glass; there it is known as the Sherrington-Kirkpatrick (SK) model [14]. In this model $H$ is the Hamiltonian (or energy function), the $i, j$ label spins, the $\sigma$ their orientation (up/down) and the $\{J\}$ are the exchange interactions between pairs of spins.

In the latter case one was naturally interested in the effects of temperature and of phase transitions as it is varied. In the standard procedure of Gibbsian statistical mechanics, in thermal equilibrium the probability of a microstate $\{\sigma\}$ is given by

---

[1] Spin glasses were originally observed as magnetic alloys with unusual non-periodic spin ordering. They were also later recognized as having many other fascinating glassy properties.

[2] This restriction is not essential but represents the potentially hardest case.

Physics and Complexity: An Introduction

$$\mathscr{P}(\{\sigma\}) = Z^{-1} \exp[-H_{\{J_{ij}\}}(\{\sigma\})/T] \tag{2}$$

where Z is the partition function

$$Z = \sum_{\{\sigma\}} \exp[-H_{\{J_{ij}\}}(\{\sigma\})/T]; \tag{3}$$

the subscript $\{J_{ij}\}$ has been added to $H$ to make explicit that it is for the particular instance of the (random) choice of $\{J_{ij}\}$. Again in the spirit of statistical physics one may usefully consider typical physical properties over realizations of the quenched disorder, obtainable by averaging them over those choices.[3]

Solving the SK model has been a great challenge and has led to new and subtle mathematical techniques and theoretical conceptualizations, backed by new computer simulational methodologies and experimentation, the detailed discussion of which is beyond the scope of this brief report. However a brief sketch will be given of some of the conceptual deductions.

Let us start pictorially. A cartoon of the situation is that of a hierarchically rugged landscape to describe the energy/cost as a function of position in the space of microscopic coordinates and such that for any local perturbations of the microscopic state that allow only downhill moves the system rapidly gets stuck and it is impossible to iterate to the true minimum or even a state close to it. Adding temperature allows also uphill moves with a probability related to $\exp[-\delta H/T]$ where $\delta H$ is the energy change. But still for $T < T_g$ the system has this glassy hindrance to equilibration, a non-ergodicity that shows up, for example, in differences in response functions measured with different historical protocols.

Theoretical studies of the SK model have given this picture substance, clarification and quantification, partly by introduction of new concepts beyond those of conventional statistical physics, especially through the work of Giorgio Parisi [12, 13].

Let us assume that, at any temperature of interest, our system has possibly several essentially separate macrostates, which we label by indices $\{S\}$. A useful measure of similarity of two macrostates $S$, $S'$ is given by their 'overlap', defined as

$$q_{SS'} = N^{-1} \sum_i \langle \sigma_i \rangle_S \langle \sigma_i \rangle_{S'}. \tag{4}$$

where $\langle \sigma_i \rangle_S$ measures the thermal average of $\sigma_i$ in macrostate $S$.

The distribution of overlaps is given by

$$P_{\{J_{ij}\}}(q) = \sum_{S,S'} W_S W'_S \delta(q - q_{SS'}), \tag{5}$$

where $W_S$ is the probability of finding the system in macrostate $S$.

---

[3] This is in contrast with traditional computer science which has been more concerned with worst instances.

In general, the macrostates can depend on the specific choice of the $\{J_{ij}\}$ but for the SK model the average of $P_{\{J_{ij}\}}(q)$ can be calculated, as also other more complicated distributions of the $q_{SS'}$, such as the correlation of pairwise overlap distributions for 3 macrostates $S$, $S'$, $S''$.

For a simple (non-complex) system there is only one thermodynamically relevant macrostate and hence $\overline{P(q)}$ has a single delta function peak; at $q = 0$ for a paramagnet (in the absence of an external field) and at $q = m^2$ for a ferromagnet, where $m$ is the magnetization per spin. In contrast, in a complex system $\overline{P(q)}$ has structure indicating many relevant macrostates.[4] This is the case for the SK model beneath a critical temperature and for sufficient frustration, as measured by the ratio of the standard deviation of $P(J)$ compared with its mean. Furthemore, other measures of the $q$-distribution indicate a hierarchical structure, ultrametricity and a phylogenic-tree structure for relating overlaps of macrostates, chaotic evolution with variations of global parameters, and also non-self-averaging of appropriate measures.

These observations and others give substance to and quantify the rugged landscape picture with macrostate barriers impenetrable on timescales becoming infinite with $N$. For finite-ranged spin glasses this picture must be relaxed to have only finite barriers, but still with a non-trivial phase transition to a glassy state.

The macroscopic dynamics in the spin glass phase also shows novel and interesting glassy behaviour,[5] never equilibrating and having significant deviations from the usual fluctuation-dissipation relationship.[6]

A brief introduction to the methodolgies to arrive at these conclusions is deferred to a later section.

## 3 Transfers and Extensions

The knowledge gained from such spin glass studies has been applied to increasing understanding of several other physically different systems and problems, via mathematical and conceptual transfers and extensions. Conversely these other systems have presented interesting new challenges for statistical physics. In this section we shall illustrate this briefly via discussion of some of these transfers and stimulating extensions.

In static/thermodynamic extensions there exist several different analogues of the quenched and annealed microscopic variables, $\{J\}$ and $\{\sigma\}$ above, and of the intensive controls, such as $T$. Naturally, in dynamics of systems with quenched disorder the

---

[4] The overline indicates an average over the quenched disorder.

[5] There are several possible microscopic dynamics that leads to the same equilibrium/Gibbsian measure, but all such employing local dynamics lead to glassiness.

[6] Instead one finds a modified fluctuation-dissipation relation with the temperature normalized by the instantaneous auto-correlation.

Physics and Complexity: An Introduction                                    123

annealed variables (such as the $\{\sigma\}$ above) become dynamical, but also one can consider cases in which the previously quenched parameters are also dynamical but with slower fundamental microscopic timescales.[7]


## 3.1 Optimization and satisfiability

Already in Sect. 2, one example of Eq. (1) as an optimization problem was given (the Dean's Problem). Another classic hard computer science optimization problem is that of equipartitioning a random graph so as to minimise the cross-links. In this case the cost function to minimise can be again be written as in Eq. (1), now with the $\{i\}$ labelling vertices of the graph, the $\{J_{ij}\}$ equal to 1 on edges/graph-links between vertices and zero where there is no link between $i$ and $j$, the $\{\sigma_i = \pm\}$ indicating whether vertices $i$ are in the first or second partition and with the frustrating constraint $\sum_i \sigma_i = 0$ imposing equipartitioning. Without the global constraint this is a random ferromagnet, but with it the system is in the same complexity class as a spin glass.

Another classic hard optimization problem that extends Eq. (1) in an apparently simple way but in fact leads to new consequence is that of random $K$-satisfiability ($K$-SAT) [11]. Here the object is to investigate the simultaneous satisfiability of many, $M$, randomly chosen clauses, each made up of $K$ possible microscopic conditions involving a large number, $N$, of binary variables. Labelling the variables $\{\sigma_i\} = \{\pm 1\}$ and writing $x_i$ to indicate $\sigma_i = 1$ and $\overline{x_i}$ to indicate $\sigma_i = -1$, a $K$-clause has the form

$$(y_{i_1} \text{ or } y_{i_2} \text{ or} \ldots y_{i_K}); \quad i = 1, \ldots M \tag{6}$$

where the $y_{i_j}$ are $x_{i_j}$ or $\overline{x_{i_j}}$. In Random K-Sat the $\{i_j\}$ are chosen randomly from the $N$ possibilities and the choice of $y_{i_j} = x_{i_j}$ or $\overline{x_{i_j}}$ is also random, in both cases then quenched. In this case one finds, for the thermodynamically relevant typical system, that there are two transitions as the ratio $\alpha = M/N$ is increased in the limit $N \to \infty$; for $\alpha > \alpha_{c1}$ it is not possible to satify all the clauses simultaneously (UNSAT), for $\alpha < \alpha_{c1}$ the problem is satisfiable in principle (SAT), but for $\alpha_{c2} < \alpha < \alpha_{c1}$ it is very difficult to satisfy (in the sense that all simple local variational algorithms stick); this region is known as HARD-SAT. These distinctions are attributable to regions of fundamentally different fractionation of the space of satisfiability, different levels of complexity.

---

[7] Sometimes one speaks of fast and slow microscopic variables but it should be emphasised that these refer to the underlying microscopic time-scales. Glassiness leads to much slower macroscopic timescales.

## 3.2 K-Spin Glass

In fact, again there was a stimulating precursor of this $K$-SAT discovery in a *"what-if"* extension of the SK model [3, 4] in which the 2-spin interactions of Eq. (1) are replaced by K-spin interactions:

$$H_K = - \sum_{(i_1, i_2 \ldots i_K)} J_{i_1, i_2 \ldots i_K} \sigma_{i_1} \sigma_{i_2} \ldots \sigma_{i_K} \tag{7}$$

in which the $J_{i_1, i_2 \ldots i_K}$ are again chosen randomly and independently from an intensive distribution of zero mean. In this case, two different phase transitions are observed as a function of temperature, a lower thermodynamic transition and a dynamical transition that is at a slightly higher temperature, both to complex spin glass phases. The thermodynamic transition represents what is achievable in principle in a situation in which all microstates can be accessed; the dynamical transition represents the situation where the system gets stuck and cannot explore all the possibilities, analogues of HARDSAT-UNSAT and SAT-HARDSAT.

The $K$-spin glass is also complex with a non-trivial overlap distribution function $P(q)$ but now the state first reached as the transitions are crossed has a different structure from that found for the 2-spin case. Now

$$\overline{P(q)} = (1 - x)\delta(q - q_{min}) + x\delta(q - q_{max}); \tag{8}$$

in contrast with the SK case where there is continuous weight below the maximum $q_{max}$. The two delta functions demonstrate that there is still the complexity of many equivalent but different macrostates, but now with equal mutual orthogonalities (as compared with the 2-spin SK case where there is a continuous range of overlaps of the macrostates). This situation turns out to be quite common in many extensions beyond SK.

## 3.3 Statics, Dynamics and Temperature

At this point it is perhaps useful to say a few more words about the differences between statics/thermodynamics and dynamics in statistical physics, and about types of micro-dynamics and analogues of temperature.

In a physical system one often wishes to study thermodynamic equilibrium, assuming all microstates are attainable if one waits long enough. In optimization problems one typically has two types of problem; the first determining what is attainable in principle, the second considering how to attain it. The former is the analogue of thermodynamic equilibrium, the latter of dynamics.

# Physics and Complexity: An Introduction

In a physical system the true microscopic dynamics is given by nature. However, in optimization studies the investigator has the opportunity to determine the micro-dynamics through the computer algorithms he or she chooses to employ .

Temperature enters the statistical mechanics of a physical problem in the standard Boltzmann-Gibbs ensemble fashion, or as a measure of the stochastic noise in the dynamics. We have already noted that it can also enter an optimization problem in a very similar fashion if there is inbuilt uncertainty in the quantity to be optimized. But stochastic noise can also usefully be introduced into the artificial computer algorith-mic dynamics used to try to find that optimum. This is the basis of the optimization technique of simulated annealing where noise of variance $T_A$ is deliberately intro-duced to enable the probabalistic scaling of barriers, and then gradually reduced to zero [10].

## 3.4 Neural Networks

The brain is made up of a very large number of neurons, firing at different rates and extents, interconnected by an even much larger number of synapses, both excitory and inhibitory. In a simple model due to Hopfield [7] one can consider a cartoon describable again by a control function of the form of Eq. (1). In this model the neu-rons $\{i\}$ are idealised by binary McCullouch-Pitts variables $\{\sigma_i = \pm 1\}$, the synapses by $\{J_{ij}\}$, positive for excitatory and negative for inhibitory, with stochastic neural microdynamics of effective temperature $T_{\text{neural}}$ emulating the width of the sigmoidal response of a neuron's output to the combined input from all its afferent synapses, weighted by the corresponding activity of the afferent neurons.

The synapses are distributed over both signs, yielding frustration, and apparently random at first sight. However actually they are coded to enable attractor basins related to memorized patterns of the neural microstates $\{\xi_i^\mu\}$; $\mu = 1, \ldots, p = \alpha N$. The similarity of a neural microstate to a pattern $\mu$ is given by an overlap

$$m^\mu = N^{-1} \sum_i \langle \sigma_i \rangle s \xi_i^\mu. \tag{9}$$

Retrieval of memory $\mu$ is the attractor process in which a system started with a small $m^\mu$ iterates towards a large value of $m^\mu$.

In Hopfield's original model he took the $\{J_{ij}\}$ to be given by the Hebb-inspired form

$$J_{ij} = p^{-1} \sum_\mu \xi_i^\mu \xi_j^\mu \tag{10}$$

with randomly quenched $\{\xi_i\}$.[8] For $\alpha$ less than a $T_{neural}$-dependent critical value $\alpha_c(T_{neural})$ patterns can be retrieved. Beyond it only quasi-random spin glass minima

---

[8] i.e. uncorrelated patterns.

126         D. Sherrington

unrelated to the memorised patterns remain ( and still only for $T_{neural}$ not too large). However, other $\{J_{ij}\}$ permit a slightly larger capacity (as also can occur for correlated patterns).

Again the landscape cartoon is illustratively useful. It can be envisaged as one for $H_{\{J_{ij}\}}$ as a function of the neural microstates (of all the neurons), with the dynamics one of motion in that landscape, searching for minima using local deviation attempts. The memory basins are large minima. Clearly one would like to have many different retrievable memories. Hence frustration is necessary. But equally, too much frustration would lead to a spin-glass like state with minima unrelated to learned memories.

This cartoon also leads immediately to the recognition that learning involves modifying the landscape so as to place the attractor minima around the states to be retrieved. This extension can be modelled minimally via a system of coupled dynamics of neurons whose state dynamics is fast (attempting retrieval or generalization) and synapses that also vary dynamically but on a much slower timescale and in response to external perturbations (yielding learning).

## 3.5 Minority Game

More examples of many-body systems with complex macrobehaviour are to be found in social systems, in which the microscopic units are people (or groups of people or institutions), sometimes co-operating, often competing. Here explicit discussion will be restricted to one simple model, the Minority Game [1], devised to emulate some features of a stockmarket. $N$ 'agents' play a game in which at each time-step each agent makes one of two choices with the objective to make the choice which is in the minority.[9] They have no direct knowledge of one another but (in the original version) make their choices based on the commonly-available knowledge of the historical actual minority choices, using their own individual stategies and experience to make their own decisions. In the spirit of minimalism we consider all agents (i) to have the same 'memories', of the minority choices for the last $m$ time-steps, (ii) to each have two strategies given by randomly chosen and quenched Boolean operators that, acting on the $m$-string of binary entries representing the minority choices for the last $m$ steps, output a binary instruction on the choice to make, (iii) using a personal 'point-score' to keep tally of how their strategies would have performed if used, increasing the score each time they would have chosen the actual minority, and (iv) using their strategy with the larger point-score. Frustration is represented in the minority requirement, while quenched disorder arises in the random choice of individual strategies.

Simulational studies of the 'volatility', the standard deviation of the actual minority choice, shows (i) a deviation from individually random choices, indicating correlation through the common information, (ii) a cusp-minimum at a critical value $\alpha_c$ of

---

[9] The philosophy is that one gets the best price by selling when most want to buy or buying when most want to sell.

Physics and Complexity: An Introduction

the ratio of the information dimension to the number of agents $\alpha = D/N = 2^m/N$, suggesting a phase transition at $\alpha_c$, (iii) ergodicity for $\alpha > \alpha_c$ but non-ergodic dependence on the point-score initialization for $\alpha < \alpha_c$, indicating that the transition represents the onset of complexity. This is reminiscent of the cusp and the ergodic-nonergodic transition observed in the susceptibilities of spin glass systems as the temperature is reduced through the spin glass transition.

Furthermore, this behaviour is essentially unaltered if the 'true' history is replaced by a fictitious 'random' history at each step, with all agents being given the same false history, indicating that it principally represents a carrier for an effective interaction between the agents. Indeed, generalising to a $D$-dimensional random history information space, considering this as a vector-space and the strategies as quenched $D$-vectors of components $\{R_i^{s,\mu}\}$; $s = 1, 2$, $\mu = 1, \ldots, D$ in that space, and averaging over the stochastically random 'information', one is led to an effective control function analogous to those of Eqs. (1) and (10) with $p$ replaced by $\alpha$, now $\{\xi_i = (R_i^1 - R_i^2)/2\}$, an extra multiplicative minus sign on the right hand side of Eq. (10), and also a random-field term dependent upon the $\{\xi_i\}$ and $\{\omega_i = (R_i^1 + R_i^2)/2\}$. As noted, there is an ergodic-nonergodic transition at a crtitical $\alpha$, but now the picture is one of the $\{\xi_i\}$ as repellers rather than the attractors of the Hopfield model.[10]

The typical behaviour of this system, as for the spin glasses, can be studied using a dynamical generating functional method [2], averaged over the choice of quenched strategies, in a manner outlined below. The averaged many-body system can then be mapped into an effective *representative agent ensemble* with *memory* and *coloured noise*, with both the noise correlations and the memory kernel determined self-consistently over the ensemble. Note that this is in contrast to (and corrects) the common assumption of a single deterministic representative agent. The phase transition from ergodic to non-ergodic is manifest by a singularity in the two-time point-sign correlation function

$$C(t, t') = N^{-1} \sum_i \overline{\mathrm{sgn}(p_i(t)\mathrm{sgn}(p_i(t')} = \langle \mathrm{sgn}(p(t)\mathrm{sgn}(p(t'))\rangle_{ens} \qquad (11)$$

where the first equality refers to the many-body problem and the second its equivalence in the effective agent ensemble.

# 4 Methodologies

For systems in equilibrium, physical observables are given by $\ln Z$ evaluated for the specific instance of any quenched parameters, or strictly the generalized generating function $\ln Z(\{\lambda\})$ where the $\{\lambda\}$ are generating fields to be taken to zero after an

---

[10] One can make the model even more minimal by allowing each agent only one strategy $\{\xi_i\}$ which (s)he either follows if its point-score is positive or acts oppositely to if the point-score is negative. This removes the random-field term and also the cusp in the *tabula rasa* volatility, but retains the ergodic-nonergodic transition [5].

appropriate operation (such as $\partial/\partial\lambda$) is performed. Hence the average over quenched disorder is given by $\overline{\ln Z}$. One would like to perform the average over quenched disorder explicitly to yield an effective system. However, since $Z$ is a sum over exponentials of a function of the variables, $\ln Z$ is difficult to average directly so instead one uses the relation

$$\ln Z = Lim_{n \to 0} \, n^{-1}(Z^n - 1) \tag{12}$$

and interprets the $Z^n$ as corresponding to a system whose variables have extra 'replica' labels, $\alpha = 1, ...n$, for which one can then average the partition function, an easier operation, at the price of needing to take the eventual limit $n \to 0$. The relevant 'order parameters' are then correlations between replicas

$$q^{\alpha\beta} = N^{-1} \sum_i \langle \sigma_i^\alpha \sigma_i^\beta \rangle_T \tag{13}$$

where $\langle .. \rangle_T$ refers to a thermal average in the effective post-averaging system. This order parameter is non-zero in the presence of frozen order, but more interestingly (and subtly) also exhibits the further remarkable feature of spontaneous replica symmetry breaking, indicating complexity. After further subtleties beyond the scope of this short introduction, there emerges an order function $q(x); x \in [0.1]$ from which the average overlap function is obtained by

$$\overline{P(q)} = dx/dq \tag{14}$$

For dynamics the analogue of the partition function $Z$ is a generating functional, which may be written symbolically as

$$Z_{dyn} = \int \prod_{\text{all variables, all times}} \delta(\text{microscopic eqns. of motion}) \exp(\{\lambda\phi\}) \tag{15}$$

where the $\phi$ symbolize the microscopic variables and a Jacobian is implicit. Averaging over the quenched disorder now induces interaction between epochs and integrating out the microscopic variables results in the effective single agent ensemble formulation, as well as emergent correlation and response functions as the dynamic order parameter analogues of the static inter-replica overlaps, exhibiting non-analyticity at a phase transition to non-ergodicity.

## 5 Conclusion

A brief illustration has been presented of how complex co-operative behaviour arises in many body systems due to the combination of frustration and disorder in the microscopics of even very simply formulated problems with very few parameters. Such systems are not only examples of Anderson's famous quotation *"More is different"* but also demonstrate that *frustration and disorder in microscopics can lead to complexity in macroscopics*; i.e. many and complexly related *different*s. Furthermore, this complexity arises in systems with very simple few-valued microscopic parameters; *complexity is **not** the same as complication* and does not require it.

There has also been demonstrated valuable transfers between systems that appear very different at first sight, through the media of mathematical modelling, conceptualization and investigatory methodologies, a situation reminiscent of the successful use of the Rosetta stone in learning an unknown language script by comparison with another that carries the same message in a different format.

The perspective taken has been of statistical physics, but it must be emphasised that the stimulation has been both from and to physics, since many of these complex systems have interesting features in their microscopic underpinning that are richer than those in the physics of conventional dictionary definition and provide new challenges to the physicist.

Also of note is how a *blue skies* attempt to understand some obscure magnetic alloys through soluble but, for the experimental alloys, unphysical modelling has led to an explosion of appreciation of new concepts, understanding and application of ideas and methologies throughout an extremely wide range of the sciences.

**Acknowledgments** The author thanks the Leverhulme Trust for the award of an Emeritus Fellowship and the UK EPSRC, the EU and the ESF for support over many years during the development of the work reported here. He also thanks many colleagues throughout the world for collaborations and valuable discussions; most of their names are given in the last slide of his 2010 Blaise Pascal lecture that can be found at [9].

# References

1. D. Challet, M. Marsili, Y.-C. Zhang, *Minority Games: Interacting Agents in Financial Markets* (Oxford University Press, Oxford, 2004)
2. A.C.C. Coolen, *Mathematical Theory of the Minority Game* (Oxford University Press, Oxford, 2005)
3. A. Crisanti, H.-J. Sommers, Z. Physik **87**, 341 (1992)
4. A. Crisanti, H.-J. Sommers, H. Horner, Z. Physik **92**, 257 (1993)
5. T. Galla, D. Sherrington, Eur. Phys. J. B **46**, 153 (2005)
6. M.R. Garey, D.S. Johnson, *Computers and Intractability: A Guide to the Theory of NP-Completeness* (W.H.Freeman, San Francisco, 1979)
7. J.J. Hopfield, Proc. Nat. Acad. Sci. **79**, 2554 (1982)
8. http://www.claymath.org/millennium/P_vs_NP/
9. http://www.eurasc.org/Galeries/AG_2010/lectures_2010.asp
10. S. Kirkpatrick, C.D. Gelatt, M.P. Vecchi, Science **220**, 671 (1983)
11. S. Kirkpatrick, B. Selman, Science **1297**, 812 (1994)
12. M. Mézard, G. Parisi, M.A. Virasoro, *Spin Glass Theory and Beyond* (World-Scientific, Singapore, 1987)
13. G. Parisi, in *Stealing the Gold*, ed. by P.M. Goldbart, N. Goldenfeld, D. Sherrington (Oxford University Press, Oxford, 2005), p. 192
14. D. Sherrington, S. Kirkpatrick, Phys. Rev. Lett. **35**, 1792 (1975)

# The Language of Systems Biology

**Marcello Delitala and Thomas Hillen**

**Abstract** Systems Biology is an interdisciplinary approach to understand biological processes that act on different scales. For example biochemical pathways steer internal cell dynamics, which can lead to cell movement. Cell movement can lead to cancer invasion and cancer invasion can lead to a disease that affects the whole body. To understand such a process, a multiscale approach is needed which can bridge the scales while retaining the complexity of the biological system. This approach is available in applied mathematics where multiscale methods have a long history. The language of Systems Biology is mathematics and it is on us to make use of these exciting mathematical methods to help to understand biological systems.

**Keywords** System biology · Multiscale modelling · Interdisciplinarity · Biological complexity

The first Kepler workshop in Heidelberg, May 16–20, 2011, has attracted a colorful mix of presentations from biologists, system biologists and mathematicians. The very interesting presentations were followed by round table discussions. It was curious to observe that eventually each discussion would revolve around "systems biology". Some people would proudly claim that they are system biologists, while others were hesitant to be associated with systems biology.

Many mathematicians consider systems biology to be just another name for mathematical modelling of biological systems. But there is more to it: systems biology can be seen with the eyes of a historian. In early education, biologists enjoy a basic

---

M. Delitala (✉)
Department of Mathematical Sciences, Politecnico di Torino, Corso Duca degli Abruzzi 24, 10129 Torino, Italy
e-mail: marcello.delitala@polito.it

T. Hillen
Centre for Mathematical Biology, University of Alberta, Edmonton, ABT6G2G1, Canada
e-mail: thillen@ualberta.ca

M. Delitala and G. Ajmone Marsan (eds.), *Managing Complexity, Reducing Perplexity*, Springer Proceedings in Mathematics & Statistics 67, DOI: 10.1007/978-3-319-03759-2_14, © Springer International Publishing Switzerland 2014

curriculum in general biology. However, very quickly, they specialize into all kind of biological fields, fragmented in several sub-disciplines. They become geneticists, or molecular biologists, or cell physiologists, or zoologists, or botanists, or ecologists etc. Each of these groups represents a certain natural scale, from molecules to genes to organisms to animals to ecosystems. Each area has its own methods and techniques and, very often, the borders between those areas are strict and interactions are quite limited.

Several successes have been achieved in each of those fields, largely increasing the knowledge and the understanding of many biological areas: nevertheless, the integration and interpretation of data is still not sufficient to understand and catch the global nature of the system. The complexity of the processes and properties of the whole cannot be simply understood by diagrams of their mutual interconnections.

Now, the rapid development of genetics in the last 20 years caused a dilemma. We all know that genes influence everything, the behavior of cells, the phenotype of individuals and the interactions within an ecosystem. Gene sequencing data become readily available and we want to benefit from them. How do the genes influence the cells, the individuals and the populations? These questions leave the area of genetics and require knowledge and expertise in these other areas of biology. A geneticist is not a cell physiologist or zoologist. The question is: How to bridge the scales from genes to cells, individuals and populations?

Thus there is the need for a multiscale modelling able to integrate and cross information from different scale layers. Biological systems are hierarchically organized with feedback and influences up and down the scales (both top down and bottom up).

The need arose for an area of biology that can connect these different areas and facilitate exchange between their methods. This is what is now called "systems biology". A definition, which most system biologists adhere to, is the science that investigates the interaction of systems that act on various scales. For many biologists this new understanding must have been like a revelation. The borders disappear and unthinkable opportunities open up: this created the boom which we experience now.

The large comprehensive data bases, made available from new experimental techniques and progresses in molecular biology, stimulated new hypothesis and experiments, demonstrating, after the undoubtable successes, the limit of classical reductionism. Biologist need to investigate relationships and complex structures with the need of mathematical techniques.

What did mathematicians do in the meantime? The applied mathematicians are trained from the very beginning to deal with multiple scales. When they learn about the diffusion equation, then they learn that the diffusion equation is not only applied to diffusion of molecules in solution, it is also used to model heat transport, cell movement, cancer invasion, population dynamics, epidemic spread, and even spread of genes in a population. They feel never restricted by scales and for instance the areas of perturbation analysis, multiscale methods, homogenization, mean field approaches etc. are widely used from applied mathematicians.

Hence, on the one hand, we see biologists who encounter a scientific revolution, and on the other hand we have mathematicians who say "we told you so—long ago!". Of course, now biologists become suspicious as if mathematicians have everything figured out—of course, they have not. But indeed, mathematics does have the right tools available. Multiscale methods are needed to bridge scale and mathematics is and will be the language of systems biology. In this context, it is evident that much hope is projected in this new approach of systems biology. It appears as the key to connect scales and finally to understand whole organisms.

In this direction, investigating the interactions of systems that act on various scales emerges as a powerful approach of the research in life sciences for a deeper insight into a complex world. System biology may help biologists to validate their mental models, exploring new pathways and dynamics: in general, system biology is necessary to achieve a deeper understanding of the biological processes and consequently a better control and prediction.

To benefit most from this development, we must give up the idea that one person can do everything, and involve ourselves in close collaboration between Biologists, Computer Scientists and Mathematicians. Fragmentation of sciences lead biologists to be trained with little mathematical tools, as well as mathematicians are not trained in biology. Thus there is the need of an interdisciplinary approach to tackle the complexity of the biological systems by establishing a common protocol and language between researchers from different areas.

In general, a system-type of approach is needed to deal with complex systems where the overall behaviour is not explained by its constituent only (the sum is more than its parts). As already discussed in the Preface of this Volume, an increasing number of applications shows these "Complexity" features: then the mathematical methods and tools developed in one specific research field may suggest and inspire new paths in other disciplines, seemingly far way from each other.

Mathematics is the language of systems biology and communication between scientists is its soul.

**Acknowledgments** M. Delitala was supported by the FIRB project - RBID08PP3J. T. Hillen was supported by NSERC.